SCIENCE

SCIENCE
PROCESS AND DISCOVERY

DENNIS FIELD

Addison-Wesley Publishers

CANADIAN CATALOGUING IN PUBLICATION DATA

Field, Dennis
Science, process and discovery

Includes index.
ISBN 0-201-18628-4

1. Science — Methodology — Case studies. I. Title.

Q175.F54 1985 501 C85-098010-0

EDITORIAL: Ron Doleman
DESIGN: Brant Cowie/Artplus Limited
TYPESETTING: Compeer Typographic Services Limited

ISBN 0-201-18628-4

Printed and bound in Canada by The Alger Press

B C D E ALG 93 92 91 90 89

TABLE OF CONTENTS

ACKNOWLEDGEMENTS

The inspiration for this book came from Dr W. C. Van Deventer, of Western Michigan University, Kalamazoo, Michigan, U.S.A. It was Dr Van Deventer who first introduced the idea of the *Principles* to the author as a way of unifying many aspects of the scientific endeavour, and who first encouraged him to write a book. Dr Van Deventer also made many excellent suggestions for improving the first draft of the manuscript.

The author would like to thank the following individuals, all of whom spent time patiently explaining things to him:

Dr John Adkins, Cavendish Laboratory, University of Cambridge, England.

Dr Frank T. Awbrey, San Diego State University, San Diego, U.S.A.

Dr Kenneth Cumming, Christian Heritage College, El Cajon, California, U.S.A.

Dr Alan Donelson, Traffic Injury Research Foundation of Canada, Ottawa, Canada.

Dr Norman Gentner, Atomic Energy of Canada Limited, Chalk River, Canada.

Dr Antony Hewish, Cavendish Laboratory, University of Cambridge, England.

Dr Ronald M. Lloyd, Pembroke, Canada.

Captain Wayne Thompson, National Defence Headquarters, Ottawa, Canada.

Dr Bill Thwaites, San Diego State University, San Diego, U.S.A.

Dr Irene Uchida, McMaster University, Hamilton, Canada.

Thanks are due to four science teachers who reviewed early drafts of the book and made many suggestions for revisions. They have improved the book enormously. These people are:

Marguerite Bryson of Delta, British Columbia.

Chuck Cohen of Victoria Park Secondary School in North York.

Joe Hircock of the Durham Board of Education.

David Lunn of Beaverbrook Secondary School in Calgary.

Don Norman of the Department of Education in Prince Edward Island.

The author would also like to thank his students, and those of his colleagues who suffered through early drafts of many of the chapters of this book. Along with the reviewers, they are responsible for modifying many incomprehensible sections and eliminating a number of errors. The ones that remain are the author's sole responsibility.

Most of the research for this book was conducted, and the first draft written, during a semester's sabbatical granted by the author's employer, the Renfrew County Board of Education, Pembroke, Ontario. The author is profoundly grateful for this opportunity, and hopes that the book's use will make this sabbatical a worthwhile investment by the Board.

PART ONE

GENERAL SCIENCE—PRINCIPLES

Isaac Newton *(1642–1727) English physical scientist and mathematician, considered by many to be one of the greatest figures in the history of science.*

1 The Cycle of Proof

The scientists you will be reading about in this book have studied many different things. Some have been astronomers. Others have worked on nuclear energy. Others again have studied the living world of plants and animals. Another group, whom we call applied scientists, have studied practical questions such as how to scare birds away from airports, or how to cure people of diseases. Some of the scientists lived hundreds of years ago— in one case thousands of years ago. Others are working right now, and the author has been able to talk to them about their work.

All these people have one thing in common. They would recognize each others' way of looking at the world and would want to find out more about it. In one form or another, they all use a method often called the scientific method. In this book, it is called the *cycle of proof*.

In this chapter, you will use the cycle of proof to look at a scientific horror story: the use of the chemical dichlorodiphenyltrichloroethane, otherwise known as DDT. First, read the story straight through. Then look at the commentary which comes afterwards. This commentary refers to the sections of the story indicated by the numbers (superscripts, like this[1]). You will see how the scientific research described in the story fits the *cycle of proof*. Most of the other sections of this book can be analyzed in the same way — but that will be a job for you to do.

A chemical of doubtful benefit

The chemical DDT was first brought into use in 1942. It was an immediate best-seller, because there was nothing like it for killing insects like the potato beetle, the spruce budworm, the mosquito, and the body louse.

Fifteen years after DDT came into common use, traces of it began to be found in the flesh of many animals, including man. Some of the affected animals lived thousands of kilometres from the nearest places where DDT had been sprayed. Plankton in the ocean, polar bears in the Arctic, and penguins in the Antarctic all showed traces of DDT.

Then, in the 1960s, bird-watchers began to notice a decline in the population of some carnivorous birds, such as bald eagles, ospreys, and peregrine falcons. Also, some eggs, particularly those of the peregrine falcons, seemed to have *thinner shells than they used to. The eggs broke easily, and often failed to hatch.*[1]

The birds had DDT in their bodies, and their eggs were defective. It occurred to some biologists that these two facts might be connected. *Perhaps the DDT was damaging the eggs.*[2]

The biologists noted, too, that it was the large carnivorous birds that were affected most—exactly what would be expected if DDT was responsible for the damage to the eggs. A bird which ate DDT-contaminated insects would certainly end up with some DDT in its body tissues. But a peregrine falcon, for example, would eat many of these smaller birds during its life, along with other animals which had also been exposed to DDT. So the falcons and other large carnivorous birds would quickly come to have much more DDT in their tissues than the smaller ones would.

If DDT really did damage the eggs, reasoned the biologists, it should be possible to make the same thing happen in the laboratory. *Birds whose food contained DDT should lay eggs with thinner shells.*[3]

In the laboratory, groups of scientists *repeatedly administered small doses of DDT to groups of birds.*[4] Other groups of birds were kept in similar surroundings, except that they were not dosed with DDT. After several months, *the birds which had received the DDT began to lay eggs with thinner shells than the birds which had not.*[5]

Commentary

The numbered paragraphs in this commentary refer to the numbers in the story above.

1. The data base. The first three paragraphs of the story describe what was already known about the topic in question before the scientists began to think about what it all meant. A great deal of scientific activity is concerned with gathering this *data base*: counting birds, examining dead birds for traces of various chemicals, investigating the work of other scientists in this area, and gathering all possible related information. A scientist or research team will always find out as much as possible about the subject before beginning any laboratory investigations. There are two main reasons for this, both of which apply to the DDT example we are using here. First, it avoids wasting time and money repeating things that have already

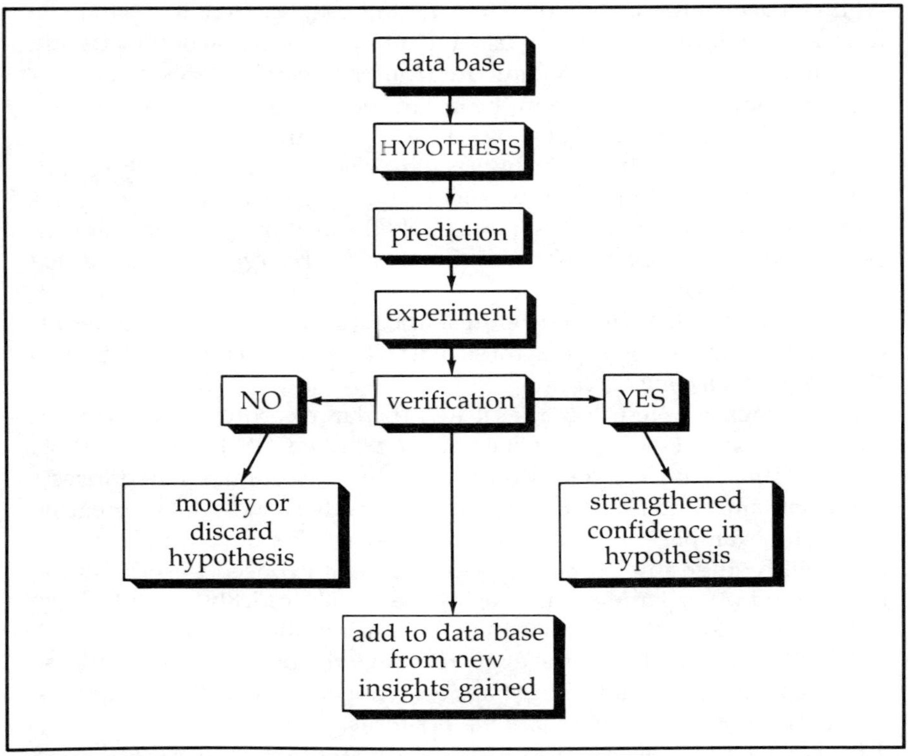

FIGURE 1-1 The Cycle of Proof

been tried. Second, examining the facts may suggest the kind of experiment which is needed. In the DDT investigations, a knowledge of the facts about DDT went a long way towards suggesting the kind of experiment that needed doing.

*2. **The hypothesis.*** From the *data base,* the biologists who were planning the investigation arrived at a possible or tentative explanation for what they knew. The explanation, in fact, turned out to be correct. Even if it hadn't, though, it would still have provided a basis for the experiment that was being planned. The tentative explanation is called the *hypothesis*. Hypotheses can usually be phrased using the word "perhaps": perhaps the DDT was damaging the eggs.

Often, when scientists arrive at a hypothesis, they realize that other facts that they already know make it more likely to be correct. In the DDT example, the biologists knew that the large carnivorous birds were affected most. This fact fit the hypothesis very well. It helps, of course, to have a *data base* which contains as much information as possible.

*3. **The prediction.*** From the hypothesis, the scientists *predict* the outcome of an experiment which they have the expertise and the resources —

equipment, time, manpower — to do. In our example, the reasoning went this way: *if* DDT is harming the eggs, *then* birds fed with DDT should start to lay defective eggs.

4. The experiment. It may seem odd that the prediction in the cycle of proof comes ahead of the experiment. In fact, both will usually be in the scientist's mind at the same time. The *experiment*, though, usually takes time and care to design. How many birds should be fed with DDT? How large should the control group be (those not fed with DDT)? How much DDT should be administered? Suitable birds have to be caught or purchased, cages have to be made ready, arrangements have to be made to feed the birds. The measurements eventually to be made on the birds' eggs have to be decided upon. Only after everything has been thought through and all the arrangements have been made can the *experiment* itself actually begin.

5. Verification. To verify something means to find out whether it is true or not. In science, the results of an *experiment* usually show whether a *prediction* turns out correctly or not. If, as in our example, things turn out much as had been expected, the scientists come away with *strengthened confidence in the hypothesis*. They still recognize that it might not be right, but their work has increased the likelihood that it is.

Often, though, a *prediction* turns out to be wrong. The results of the *experiment* are different from what was expected. If this happens, the scientists will look closely at the way the *experiment* was set up and carried out. In the end, they have to be prepared to *modify or discard the hypothesis*. An experiment on DDT might result in a conclusion such as: "we still think that DDT harms birds, but more DDT is needed than we used in our experiment." This would have been an example of a *hypothesis* being *modified*. But further experiments might have shown that DDT did not cause the egg shells to become thinner. In that case, the *hypothesis* would have had to be *discarded*.

In the various sections of this book, you will find many examples of hypotheses being modified after experiments turned out differently from what people expected. You will also find examples of hypotheses — often ones which people had believed in for a long time — having to be discarded completely.

However the experiment turns out, it will be described in specialized scientific journals, and the sum total of knowledge in the area will be increased. Other scientists working in the same field will read about the work, and will therefore increase their own knowledge and understanding of the subject. Whatever the result of the experiment, the time, money, and effort that went into it are not wasted. As a result of the work, the *size of the data base* will have been *increased*.

The rest of the story

Meanwhile, other studies were showing disturbing evidence of the effects that DDT could have on humans. Like ospreys and peregrine falcons, we are large carnivores—most of us eat meat from chickens, pigs, and cattle. DDT tends to build up in the human system in the same way as in the carnivorous birds. As a result, DDT was banned in the United States and Canada in 1970. By 1982, the number of ospreys in the coastal area between New York City and Boston rose from about 100 breeding pairs to about 1000. Numbers of birds in other threatened species have also shown a heartening increase.

Not all was well, however. Research showed that some bird species continued to be affected by DDT residues, particularly pelicans on the coast of southern California, and migrant birds like the peregrine falcon which fly between North America and Central and South America. These birds still lay eggs with thin shells, and the eggs often fail to hatch. As a result, the population remains low. Biologists expect that pelicans and peregrine falcons will continue to be threatened as long as DDT remains in use in Central and South America.

QUESTIONS

Usually, when you have finished reading a chapter in this book, you will answer some of the questions at the end of the chapter. In this chapter, the first four questions are answered for you. Read these questions and answers. Then answer the rest of the questions on your own.

1. As a result of the evidence that was gathered about the harm done by DDT, what action was taken? **ANSWER: DDT was banned in the United States and Canada.**
2. This action was not intended as an *experiment* on bird populations. Biologists, however, would like to have ended the use of DDT in order to test further the *hypothesis* that it was causing the fall in population in some bird species. For the biologists, then, ending the use of DDT was one part of an *experiment*. What other things did the biologists have to do in order to complete this *experiment*? **ANSWER: The scientists had to make careful observations on the populations of carnivorous birds in the United States and Canada.**
3. What *prediction* could be made from the *hypothesis* about the result of this *experiment*? **ANSWER: The population of the carnivorous birds should increase.**
4. What was the actual result of the *experiment* for the birds who spend all their lives in the United States and Canada? **ANSWER: The population did increase.**
5. What effect does this result have on your confidence in the original *hypothesis*? (Look at the sections of the *cycle of proof* which come after the *experiment* section.)
6. The countries of Central and South America are poorer than the United States and Canada. They cannot afford either the damage which insect pests would do, or the more expensive chemical substitutes for DDT. If this were not so, what *experiment* would you suggest to see if pelicans and peregrine falcons are harmed by DDT? Be careful to describe all parts of the *experiment*, including the measurements which would have to be made.
7. What *prediction* would you make about the result of your experiment?
8. Suppose, after several years of living in a DDT-free environment in a zoo, a group of pelicans showed no DDT in their body tissues, but they continued to lay eggs with thinner and thinner shells. In terms of the *cycle of proof*, what would this result do to your confidence in the *hypothesis* described in the article?

ANALYSIS

The first five questions are done for you. Read these and answer the last five on your own. All ten of these questions ask you to say which parts of the *cycle of proof* apply to various incidents in *The rest of the story*. Select your answers from the following alternatives (give the reason for your choice):

a) a *hypothesis*;
b) part of the *data base* the scientists used in arriving at their hypothesis;
c) a *prediction* made from the hypothesis;
d) an *experiment* which was performed or could be performed to test the hypothesis;
e) the results of an experiment which would *strengthen confidence* in the hypothesis.

1. The decrease in population of many carnivorous birds was due to the chemical insecticide DDT. **HYPOTHESIS: At first, the scientists did not know what caused the decrease in population. They put forward the tentative explanation that perhaps it was the DDT.**

2. These carnivorous birds were showing traces of DDT in their tissues, and they were laying eggs with thinner shells than before. **DATA BASE: This statement shows two facts upon which the scientists based their hypothesis.**

3. DDT was banned and biologists studied the population of some of the carnivorous birds. **EXPERIMENT: This statement describes some of the things the scientists did to see if the hypothesis was correct.**

4. During the 1970s, biologists expected that the number of many carnivorous species would increase. **PREDICTION: This statement shows the results the scientists expected from their experiment—assuming that the hypothesis was correct.**

5. By 1982, the number of ospreys in the coastal area between New York and Boston increased ten times. **STRENGTHENED CONFIDENCE IN THE HYPOTHESIS: This result of the experiment is exactly what the scientists expected. The prediction, which in turn was based on the hypothesis, turned out to be correct.**

6. Biologists believe that DDT does something to birds which makes their eggs have thinner shells.

7. The number of peregrine falcons and pelicans is decreasing.

8. DDT is still used in some countries in Central and South America.

9. Some biologists would like to see the use of DDT in Central and South America stopped. They could then study the numbers of peregrine falcons and pelicans, and also the shell thickness of their eggs.

10. The most likely outcome, they think, would be that the eggshells would become thicker over a period of years, and the population of the birds would increase.

2 The Ten Principles

In this book, you will read about the Greek philosopher Eratosthenes, who accurately calculated the size of the earth—over two thousand years ago. You will read about Charles Darwin, who was the first person to give a satisfactory answer to the question: how did all the plants and animals in the world become the way they are? You will read about the work of the great astronomers, from Tycho Brahe, Kepler, Galileo, and Newton in the seventeenth century, to Antony Hewish and Thomas Gold of our own time.

All these scientists worked in different ways, and on different problems. But as you read about their work, it will become increasingly clear that they share certain attitudes, certain ways of looking at things, and certain expectations about the kinds of things they are likely to find out. The *cycle of proof* is one way of describing the way scientists go about their work. The *Ten Principles,* to be discussed in this section, deal with some other aspects of what scientists do, how they think, and the kind of world they think they are finding.

1. The Principle of Objectivity

Scientists cultivate the ability to gather and examine facts. They base their conclusions only on these facts.

When people think of science, they usually think of experiments. A scientist's conclusions are always based firmly on the facts, and a laboratory is no more than a place specifically designed for finding out these facts.

Not only are scientific conclusions based on fact; they are based on all the facts (or as many as have been discovered), and only on the facts.

Scientists are rigorously critical of their own and their colleagues' work. Sloppy thinking doesn't last long. Neither does a conclusion based on incomplete knowledge.

Every scientist is well-read in his or her field. A visit to the science section of the library in a university or research establishment will reveal more information than could possibly be carried in one person's head. Every working day, scientific journals arrive, full of new information. Yet part of every scientist's job is to keep up with the latest work in his or her own area. Without this continual enlargement of the *data base*, there would be no science. This great stress on knowing all the available facts before beginning an investigation is part of the Principle of Objectivity too.

2. The Principle of Tentativeness

Scientists do not regard their conclusions as final, but are willing to modify them if they are contradicted by new evidence.

Every scientist knows and respects the contributions made to knowledge and understanding by the scientific giants of the past, such as Galileo, Newton, Joule, Dalton, Priestley, Lavoisier, Bohr, Harvey, Darwin, and Pasteur. Yet every one of these people reached conclusions now regarded as wrong or at least incomplete. Hypotheses, theories, and laws are constantly being tested — by *every* experiment and *every* observation — from the time they are put forward. As new facts are discovered, former beliefs and hypotheses are modified or discarded.

It's not as though the original work was done carelessly — quite the opposite. English Physicist John Adkins, of the Cavendish Laboratory at Cambridge University, puts it like this: "If you're going to get anywhere in science, you have to be almost neurotic, in checking and cross-checking, to make sure you've got it right."

3. The Principle of Consistency

Scientists assume that the behaviour of the world is describable in terms of laws which have always operated in the same way, and that the world which we now see is the result of the continuous operation of these laws.

A glance at any scientific reference book will reveal numerous facts. One example might be a table of densities of various metals. This information will almost certainly be updated as more discoveries are made. More metals will be added to the list, and the densities will be determined more accurately by better experimental techniques. Nevertheless, scientists would be most surprised if the density of mercury, for example, suddenly changed from 13 600 to 11 700kg/m^3.

Scientists are also confident that the fundamental laws which describe the way in which the universe operates will also stay the same. If the Law of Universal Gravitation was true in Newton's time, it is also true today, and will remain true in the future. Our understanding of these laws may change. The Law of Conservation of Energy, for example, is now known not to hold in its original form for nuclear reactions. It has been replaced by Einstein's Law of Conservation of Mass-Energy: $E = mc^2$. But it is our understanding of the Law, not the Law itself, that has changed.

4. The Principle of Causality

Scientists believe that every phenomenon results from discoverable causes.

Confronted with a puzzling situation, a scientist always assumes that there is a reason why that situation occurs. He or she further assumes that with enough time and effort, the cause or reason can be discovered.

At one time many people, particularly in North America, suffered from a disease known as pellagra. Victims suffered from skin rashes, diarrhea, and nausea. In extreme cases, they became psychotic and even died. The disease was unknown before the discovery of America. Around 1907, however, many people began to suffer from pellagra. The disease became especially common in the southern United States. In 1915, over 10 000 people died of it.

Why did people get pellagra? For a long time, biologists and medical researchers didn't know. They always assumed, though, that something did cause it, and that it wasn't "just something that happened."

Pellagra turned out to be the result of a dietary deficiency. The people who got it always lived in corn-growing districts, and corn formed a large proportion of their diets. But corn lacks niacin, a B vitamin. When it was discovered that the disease was in fact due to a diet without enough variety, the cure was fairly obvious. The point, though, is that researchers assumed that something did cause pellagra, and that it was their job to find out what it was.

5. The Principle of Parsimony

Scientists attempt to reduce their view of the world to the simplest possible terms. They prefer explanations, theories, and hypotheses that account for as many phenomena as possible.

Parsimony is an unwillingness to spend two dollars where one would do, or to buy a new car if the old one will go for a few more months. Scientists are parsimonious, too, in the sense that they prefer to explain things in ways that are as simple as possible. They also prefer the theories which they already know, and invent new laws, principles, and theories only when it becomes clear that the old ones just won't do.

The Principle of Parsimony was first put forward by the English monk William of Occam (1300 - 1349), and is often known to this day as "Occam's Razor," the razor being an instrument to shave things down as closely as possible. Occam put it this way: "Entities must not needlessly be multiplied."

If you came down to breakfast in the morning, and discovered a broken window, a missing radio, and footprints in the garden outside the window, there would be a number of conclusions you could come to about how it all happened. Perhaps a stone dropped from an airplane and broke the window, a monkey came in through the window and took the radio, and the footprints were caused by a neighbour looking in to see what was going on. All this is possible—it fits well with the principle of causality—but it wouldn't strike you as the most likely explanation. There are four "entities": the airplane, the stone, the monkey, and the curious neighbour. The most likely explanation has only one "entity" — a burglar. The airplane, the stone, the monkey, and the neighbour are a result of Occam's "needless multiplication."

6. The Principle of Materiality

Scientists prefer material and mechanical explanations of phenomena, rather than those which depend on non-material or supernatural factors.

A primitive person, seeing someone struck and killed by lightning, might explain it as a blow struck by an angry god. The not-so-primitive Norsemen, in fact, had a god of thunder, Thor. Even today, some people might say that the man was killed as a judgement sent by God because of some sin he had committed. A scientist, seeing the same thing, would explain it by looking at the physical laws that describe the behaviour of atmospheric electricity.

The Principle of Materiality does not prevent the scientist from believing in God, Allah, Mohammed, or Thor. It does make supernatural explanations something of a last resort, though. Before a scientist proposes that something can best be explained by the action of visitors from outer space, or by the miraculous hand of God, he or she will first investigate every other possible explanation. The scientist does not say that miracles can never occur — that would be contrary to the Principle of Tentativeness. But all other possibilities have to be eliminated first, before the scientist will accept that a miracle has taken place.

7. The Principle of Relativeness

Scientists think of the world, and the phenomena in it, as consisting of sets of relationships rather than of absolutes.

Meteorology—the study of weather and the climate— provides a good example of another aspect of the kind of world scientists expect to find. We can't say that any single factor controls what the weather will be like, and that nothing else comes into it. In fact, many factors work together on any given day to determine whether it's warm and sunny or cold and rainy. The movement of the earth around the sun, the earth's daily rotation on its axis, the convection effects of rising warm air, the effects of ocean currents: all these factors and many others combine to produce strong winds, heavy clouds, or warm temperatures. The influence of each of these factors on the weather can't be considered by itself. Each can only be understood *relative* to the others. The weather we end up with is a result of the *relationship* among all these factors.

The same situation applies in physics. People used to think of two separate laws: the Law of Conservation of Mass and the Law of Conservation of Energy. Then, in the early years of this century, Albert Einstein showed that this picture was not quite correct, and that, in fact, mass and energy were two forms of the same thing. Now we talk of the Law of Conservation of Mass-Energy; one can be converted into the other, as in a nuclear bomb or a nuclear reactor. When such a conversion takes place, the amount of mass and the amount of energy are related by the equation $E = mc^2$, c being the speed of light. Rather than two absolutes, mass and energy, which can be looked at separately, we have two forms of the same thing, with a *relationship* between them.

8. The Principle of Dynamism

Scientists expect nature to be dynamic rather than static, and to show variation and change.

The weather, clearly, is something that changes from day to day, and never repeats itself exactly. We refer to something that is constantly changing as being *dynamic*. Over a longer time, too, changes in the climate occur. The branch of science known as climatology deals with these longer-term variations. Ice ages come and go, deserts take over regions that had been fertile, and even the levels of the oceans rise and fall as the polar icecaps become colder or warmer. Similarly, the plants and animals living on the land and in the sea, and the relationships among them, are in a state of constant change. The Theory of Evolution, described in detail later in this book, is an attempt to explain the reasons for this kind of change.

There's no contradiction here with the Principle of Consistency. The two principles, Consistency and Dynamism, represent the two sides of the same coin. Even though the basic factors which govern the climate or the ecology stay the same, the climate and the ecology themselves are always changing.

9. The Principle of Continuous Discovery

Scientists hope that it will be possible to go on learning about the material world, and the material universe of which it is part, until eventually all may be understood.

We have already seen, with the principles of Objectivity and Causality, that scientists are curious about the world and about the things in it. They are always eager to learn more about how things behave, and why they behave as they do. Scientists are confident that their skills and their methods will let them find out about just about anything.

There are limits to this, of course. Although scientists hope not, it may be that some things are just too complicated for anyone ever to be able to understand them. Also, it often happens that scientists who would dearly like to know more about something have to wait until the right knowledge in some other field comes along, or the right experimental techniques are discovered. Astronomers were limited in what they could find out about the heavens until the telescope was invented. Medical researchers couldn't find out much about the causes of cancer until the biochemistry of animal cells was understood in some detail. And the transformation of mass into energy in nuclear reactors had to wait until the process of nuclear fission had been discovered in the late 1930s.

These limitations aside, every scientist is convinced that his or her work will reveal things which were not known before. Otherwise, there wouldn't be much point in doing science at all.

10. The Principle of Social Limitation

The social framework within which scientists operate may determine and limit the kinds of problems with which they work, and may also influence their conclusions.

Some of the subjects for scientific research discussed in this section have obvious practical value. Medical research, weather studies, improving the safety or efficiency of nuclear reactors: all this is done because people hope it will benefit society as a whole. This type of research is called *applied* research, and is the type likely to be encouraged by governments and private companies. The eventual results are intended to be of use to people, and will, it is hoped, yield political or financial benefits for those who decide whether the research should be done. Generally, government or industry decides upon the objectives for the research, and then provides money for scientists to carry it out.

Another kind of research is necessary too. This type is called *pure* or *curiosity-based* research. In this type of research, the objective is just to find out more about something, or to understand it better. There does not have to be a practical use for the discoveries.

Often, though, curiosity-based research does lead to knowledge of great importance to society as a whole. Albert Einstein worked out the Theory of Relativity without any idea that one of the relationships that came out of his theory, $E = mc^2$, would one day be used as one of the basic foundations for nuclear bombs and nuclear reactors.

Much curiosity-based research takes place in the universities, and it might at first seem that the needs of society as a whole don't apply to it. But these needs and priorities have a way of making themselves felt even here. Money for various research projects has to come from somewhere, and much of the equipment required is very complicated and expensive. The amount of money available is controlled by private foundations of various kinds, and by governments.

Scientists' own values may influence the type of research they do. A number of physicists and chemists, for example, refused to work on the development of the hydrogen bomb.

Some scientific conclusions can be very controversial. Later in this book, you will read about the 16th century astronomer Galileo Galilei, and the trouble he got into when he came to the conclusion that the sun, not the earth, was at the centre of our part of the heavens. In our own century, the Russian geneticist Trofim Lysenko avoided getting into similar trouble with the authorities. Contrary to all the evidence, he maintained that acquired characteristics in plants and animals could be inherited — a conclusion that just happened to fit in with the Marxist political views of the Soviet government. It needs courage to go against the accepted beliefs of the time. We will never know how often a scientist might have come to a conclusion, but did not publish it because of a fear of being unpopular.

In this book, the scientific way of going about things is put forward as something to be admired, and as something which has contributed a great deal to the well-being of us all. It should not be concluded from this that all scientists are perfect, or even that they are any better than the rest of us. But perhaps what makes science successful is that scientists are taught to be critical of other people's results, and to be very careful before arriving at a conclusion.

ANALYSIS

Each of the statements in questions 1-10 provides an example of one of the *Ten Principles*. For each, state which principle is involved, and explain. The first two are done for you.

1. The girl noticed that her little brother was crying. Something must have upset him, she thought. **PRINCIPLE OF CAUSALITY: She assumed that there was some cause for what she saw.**
2. The man visited the neighbourhood he'd grown up in. The houses were the same ones, but there had been many changes. What struck him first were the trees—much bigger than when he was a boy. They made the area look completely different. **PRINCIPLE OF DYNAMISM: We expect that changes will occur over a period of time.**
3. The vet examined the cat thoroughly, and had a blood sample analyzed, before telling the cat's owner what was wrong.
4. The chemist was employed by a major food processing company. He wanted to work on an idea he had for keeping fresh fruit from rotting. But the company wouldn't let him. Their business was mostly in bread and cereals, and they weren't ready to expand into selling fruit.
5. The old lady recovered very quickly after a heart attack. Her husband thought it was a miracle, but the doctor knew that she had been very fit, and had received very prompt attention at the hospital.
6. The lady from the garden centre was explaining why the lawn hadn't grown well. The area was too shady for the kind of seed that had been used, and the soil was too acid. Another kind of seed would have done better in that shady area, and the acidity wouldn't have been a problem if the grass had been healthier.
7. The town's mayor wanted to know why the water supply had become contaminated. "I'm not sure," said the engineer. "I think I know what might be wrong, but we'll have to wait until we see the results of the new tests we've just had done."
8. We know much more now about what causes cancer than we did fifty years ago. A great deal of research is still being done, and the researchers expect to find out more and more. Perhaps in another fifty years, cancer won't be the cause of so many people's deaths.
9. The car wouldn't start, the lights wouldn't go on, the radio didn't work, and the horn wouldn't blow. The mechanic at once suspected that all these things might have been caused by a faulty battery.

10. The boy knew that eating tomatoes always made his skin break out in red spots. He was at a birthday party, and ate the tomatoes that were in the salad. Sure enough, the next day his face was covered with red spots.

3 The Outbreak of Food Poisoning: A Case Study

The nurses at the hospital were having the worst few days any of them could remember. People were coming into the emergency department in great numbers. Some had violent stomach cramps. Others had uncontrollable diarrhea or vomiting. Some unfortunate people had all three symptoms.

The Calvert family's experience was fairly typical. First, Mrs. Calvert brought in two-year-old Karen, who was in great distress about something, but was too young and in too much pain to tell anyone what it was. Twelve-year-old Michael, who had just arrived home after a couple of days spent with a friend, was with them, armed with a pile of science fiction magazines—his latest craze. While they were still in the waiting room, in came Brian, who was seven, followed by his father. Both looked very sick. Brian was holding his stomach, and wailing pathetically. While driving Brian in, Mr. Calvert had been almost doubled over with wrenching stomach cramps — it was all he could do to park the car properly in the hospital parking lot.

Mrs. Calvert was quite angry with her husband. His father had suffered quite badly with stomach ulcers, and she had always been telling him that he'd get them too. He didn't eat well, often going to work without breakfast and bolting down his lunch while he worked. He was a policeman, and lived under a great deal of stress. ''These cramps are a terrible warning,'' she thought, and she resolved to get him to change his habits as soon as this attack was over. Perhaps now at last he'd listen.

But first, Karen and Brian had to be dealt with. Brian's sickness she could understand. She was sure he and some of his friends had climbed over the fence into a neighbour's orchard, and had eaten some apples which weren't ripe yet. As for young Karen, she'd been outside earlier in the day, and as she was bringing Karen inside for her afternoon nap, Mrs.

Calvert had noticed bees nearby. ''Perhaps,'' she thought, ''a bee stung Karen and this is a delayed reaction.''

Nurse Davidson thought differently. She realized that it wasn't likely that three members of the Calvert family would be hit by three different ailments all within a couple of hours. ''This is an epidemic of something,'' she said. ''We've been frantic all day trying to look after everyone who's been coming in — all with the same kind of symptoms.'' And when Mrs. Calvert herself began to feel ill, she had to agree that Nurse Davidson was probably right about that.

Michael Calvert had just read a particularly unlikely science-fiction story, and he tried to explain another possibility to Nurse Davidson. The human digestive system had probably changed that day, he explained, so that from now on people wouldn't be able to digest vegetables and meat as they had before. In future, they'd have to eat oats and hay, like horses. Nurse Davidson thought this was a ridiculous idea. ''As if I didn't have enough to do today,'' she thought, ''without having to put up with this silly nonsense.''

As is usual in such cases, the food poisoning didn't hit everyone equally. Food poisoning is not as simple as many people think: its effects depend on the age and general health of the people affected, and for very old people, or people who are already ill, it can be a killer.

In the episode which affected the Calvert family, two old people and one three-year-old died. Two people who had been recovering from other internal problems became so ill that they had to be readmitted to hospital.

The local newspaper's correspondence columns were full of opinions about the outbreak. One letter blamed the townspeople's wickedness, and attributed the outbreak to a visitation from the Devil. The local doctors disagreed. Something was behind the outbreak, they knew, and they called upon experts from a nearby University's Department of Epidemiology to tell them what it was.

The epidemic changed from day to day. At first, only a few people were affected with mild bowel problems. A few days later, at the time that the epidemic hit the Calverts, doctors couldn't keep up with the numbers of patients, many of whom were very ill indeed. Later still, more of the patients complained of severe stomach cramps, but vomiting was less common. By the end of the week, the outbreak was beginning to die down.

The epidemiologists tracked the outbreak down to a particular poultry processor. Before coming to a final conclusion, though, they waited for the results of a few last confirmatory tests. Caution was certainly called for. Once the business responsible for the outbreak had been identified, they expected that it would have to close. That was usually what happened in these cases.

The university people had a rather different view of the outbreak than the local people. Of course, they were pleased that their expertise was useful in finding out where the epidemic was coming from, and therefore in stamping it out. But to them, the incident was more than that. It represented an excellent opportunity to find out more about outbreaks of food poisoning in general. Their job was to put the suffering and emotions to one side and to unearth as many of the facts of the situation as they could. The more the epidemiologists knew, the better they could help local officials avoid or at least control similar outbreaks in the future. In the future, they hoped, there would be fewer cases like the Calverts, where whole families were stricken with such terrible suffering.

ANALYSIS

Each of the following quotations from the article provides an example of one of the *Ten Principles*. The quotations are underlined in the chapter for reference.

For each of the quotations, say which principle is in operation, and explain. Each quotation represents a different principle, so all ten principles will be included in your answer. The first question is done for you.

1. ". . . it wasn't likely that three members of the Calvert family would be hit by three different ailments all within a couple of hours." The **PRINCIPLE OF PARSIMONY makes us suspect that there is only one cause for these people being ill.**
2. "Michael . . . tried to explain his idea to Nurse Davidson. The human digestion system had probably changed that day. . . . Nurse Davidson thought this was a ridiculous idea."
3. "Food poisoning is not as simple as many people think; its effects depend on the age and general health of the people affected."
4. "One letter . . . attributed the outbreak to a visitation from the Devil. The local doctors disagreed."
5. "Something was behind the outbreak, they knew. . . ."
6. "The epidemic changed from day to day."
7. "Before coming to a final conclusion, though, they waited for the results of a few last confirmatory tests."
8. "Caution was certainly called for. Once the business responsible for the outbreak had been identified, they expected that it would have to close."
9. "It represented an excellent opportunity to find out more about outbreaks of food poisoning in general."
10. "Their job was to put the suffering and emotions to one side and to unearth as many of the facts of the situation as they could."

PART TWO

CASES IN GENERAL SCIENCE

Galileo Galilei *(1564–1642) Italian mathematician and astronomer—the first man to use the telescope to study the skies, and to propose the idea that the earth revolved around the sun.*

4 The Case of Brenda Brown

A routine case at the end of a long week

It was Friday afternoon. There had been an epidemic of food poisoning in the town, and all the doctors were tired. Most exhausted of all, perhaps, was Dr. Tom Fisher, who combined a busy general practice with the chairmanship of the local Physicians' Association. It had been Dr. Fisher's task during the week to visit schools and local radio stations to get the message across: be meticulous in personal hygiene and cleanliness of food preparation.

Now, fourteen-year-old Brenda Brown was his last patient before a well-earned weekend break. Brenda's complaint was by now all too familiar to Dr. Fisher, but most embarrassing to Brenda. She hadn't felt well all day, and now, in the waiting room, she had been publicly and humiliatingly sick. By the time she went with her mother into the consulting room, Brenda was pale, moist with sweat and just a few tears, and scratching nervously at her hands. Dr. Fisher asked when she'd started to feel ill. The answer was just as he'd expected—he'd heard it a dozen times already that afternoon: just this morning, she told him.

At least, thought Dr. Fisher, this isn't going to take very long. The familiar reassuring words came easily to his mind: home, bed, plenty of fluids, quite all right in a few days. "Any tummy pains?" he asked. Again the answer was as he expected: "Yes. They began just before I left to come here."

A warning note

A warning note sounded in Dr. Fisher's tired brain. ''Just come here from where?'' he asked. ''Well,'' replied Brenda, ''I had to spend most of the week with my grandmother out in the country. She was sick last week, you see, and needed someone to look after her. Nobody else could do it, so I went. My mum just came to bring me home this morning.''

Brenda's answers to his next few questions showed the doctor that he'd have to rethink his first diagnosis. Brenda hadn't been anywhere near the town for the last week. Her symptoms had begun even before she'd left her grandmother's, three hours' drive away. She couldn't be part of this epidemic. He'd have to think again.

The doctor's diagnosis

Another idea occurred to him. She'd never been ill before, had never had an operation, so she must still have her appendix. Appendicitis would explain all three of the symptoms: nausea, pain, and her loss of appetite — she'd told him she hadn't felt like eating anything since Thursday morning. Appendicitis calls for prompt action: a ruptured appendix can be fatal.

Dr. Fisher began his examination. Brenda was tender in the right places around the navel. ''Surgery for you, my girl,'' he thought to himself. Before committing himself, though, he sent Brenda over to the hospital for the blood and urine tests routinely carried out before any surgery. Prospects for a pleasant weekend receded, and while he waited for the results, Dr. Fisher and his nurse tidied up some routine office paperwork.

The lab sends back the test results

The lab called back with the test results gratifyingly quickly. As the doctor had expected, the blood tests showed nothing. But the other finding was a surprise. Noticing that the urine was brown in colour, the technicians had performed a bilirubin test, and had found the telltale bilirubin grains characteristic of jaundice. Appendicitis doesn't turn the urine brown, and doesn't respond to the bilirubin test.

Dr. Fisher went to the hospital to see Brenda again. More questions, more answers. Yes, she had in fact had quite bad stomach cramps on Monday or Tuesday — she couldn't remember which. The cramps had passed quite quickly, but she had been rather tender since, and hadn't really felt like eating much all week. And yes, her urine had been rather brown. She hadn't thought to mention all this before, partly because she'd put it down to her grandmother's food, which she didn't like. The other reason

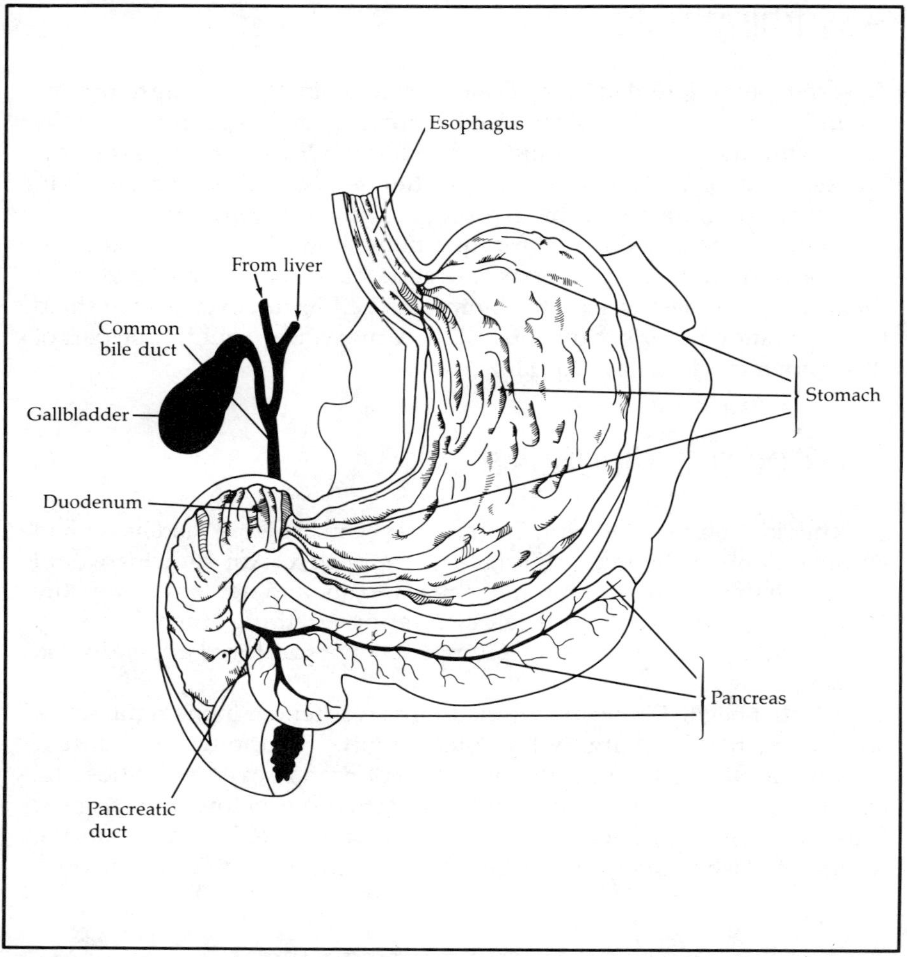

FIGURE 4-1 The Digestive System

was that it was Friday afternoon. She could see the doctor was tired, and so was she. What she'd really wanted was some pills, so she could go back home, sleep, and feel better for the rest of the weekend.

The facts behind Dr. Fisher's diagnosis

Into the *duodenum* are secreted various substances which are needed for the digestion of food in the *small intestine*. One of these is *bile*, which helps in the digestion of fats. Bile is produced by the *liver* and is stored in the *gallbladder* for use when fats come down from the *stomach*. Bile contains strong brown pigments, the most important of which is *bilirubin*.

Sometimes, small, solid lumps — *gallstones* — crystallize out from the bile. These can go down into the *bile duct*. If they are trapped there, they can cause severe pain.

If gallstones obstruct the normal flow of bile into the duodenum, the brown pigments accumulate in the blood. This causes *jaundice*—a yellowish staining of the skin and the whites of the eyes. In jaundice, the skin often becomes very itchy. The gall pigments also stain the urine brown. They can be detected in the urine by the *bilirubin* test.

Jaundice may also be caused by *hepatitis*— damage to the liver. There are three forms. *Infectious hepatitis* is transmitted through the feces, and is often found where sanitation is poor. *Serum hepatitis* is transmitted by direct contact between the blood of one person and another— for example, by a syringe. It is usually a good deal more severe than infectious hepatitis. *Toxic hepatitis* is caused by poisoning, for example, by chemicals such as phosphorus and carbon tetrachloride. These can be introduced into the body through the mouth or the lungs.

A new diagnosis

Dr. Fisher took Brenda over to the large window of the examination room. Sure enough, the whites of her eyes had that slightly yellow, jaundiced look. Usually he would have noticed by now that her skin was yellow too, but her healthy-looking tan had hidden that. Now he also remembered that she had been scratching her hands when he'd first seen her. He'd attributed that to nervousness at the time. But yes, she said, her hands did itch, and her feet too. It was becoming clearer in Dr. Fisher's mind what the trouble might be: some form of hepatitis. The next task was to find out just which kind. Brenda had to be questioned yet again.

Brenda hadn't known anyone with jaundice, and hadn't been away from home recently, except to her grandmother's. There was therefore no reason to suspect infectious hepatitis. She hadn't had any injections or blood transfusions in the last few months. No, she certainly hadn't injected any kind of drug into herself, so serum hepatitis seemed to be ruled out too. Just as well, thought the doctor. There is no good treatment for this type of hepatitis. Another of his young patients, he remembered grimly, had attempted suicide during the depression that usually accompanies the long convalescence. Finally, Brenda hadn't been near any unusual chemicals, so it probably wasn't the third type of hepatitis, toxic hepatitis.

Dr. Fisher now had enough information to arrive at a working diagnosis: obstructive jaundice, caused by an obstruction in the bile duct. Brenda was admitted to the hospital, put on intravenous feeding, and kept under close observation. Meanwhile, Dr. Fisher conferred with his surgical colleagues to schedule an operation at the beginning of the following week.

The operation

To make sure that the working diagnosis was correct, the doctor ordered more blood tests — specific ones that allow doctors to distinguish between the various types of jaundice. Knowing that something must be making the solid lumps that were blocking the common bile duct, the surgeon and Dr. Fisher ordered X-rays of this region. Indeed, a number of stones were building up in the gallbladder, and one of these had moved down the bile duct. More stones would probably be released from time to time, and Brenda would have the same problem—or worse—again. So, as well as removing the stone in the bile duct, the surgeon took out the gallbladder too.

After a most uncomfortable few days in the hospital, Brenda came home and recuperated for much of the rest of the summer. Now she has to avoid foods that are rich in fat; she doesn't have the reserve of bile to digest them that the gallbladder used to contain. As for Dr. Fisher, he missed a good part of his weekend. What a good thing though, he thought, that he hadn't been quite satisfied with his Friday afternoon diagnosis.

QUESTIONS

Questions 1 – 7 refer to the section *The facts behind Dr. Fisher's diagnosis.*

1. State the function of each organ named.
   a) small intestine
   b) liver
   c) gallbladder
   d) bile duct.
2. What are gallstones and why can they be painful?
3. How would you be able to tell if you had jaundice?
4. How do gallstones cause jaundice?
5. Brenda's problem was caused by obstructive jaundice. Name three other possible causes of jaundice. For each, state how people contract the illness.
6. What is bilirubin?
7. If a test shows bilirubin in a patient's urine, what can the doctor conclude?

ANALYSIS

Questions 1 – 7 refer to sections 1 and 2 of the narrative.

1. What was Dr. Fisher's *hypothesis* when he first saw Brenda?
2. What were some of the facts from the *data base* upon which he based his hypothesis?
3. He performed an *experiment*: asking Brenda when she'd started to feel ill. What *prediction* had he made about her answer?
4. As a result of the *experiment,* Dr. Fisher's confidence in his *hypothesis* strengthened or weakened. Explain why.
5. Performing another *experiment,* Dr. Fisher asked Brenda another question. What was the question?
6. What *prediction* did he make in his mind about the answer to the question?

7. What did Brenda say which made Dr. Fisher discard his first *hypothesis*?

The following questions refer to sections 3 and 4 of the narrative.

8. What was Dr. Fisher's second hypothesis?
9. In his examination, what *prediction* had he made to himself about the location of the pain?
10. What simple *experiment* did he do to test the prediction?
11. Next, he sent Brenda to the hospital for tests. What would have been his *prediction* about the results of these tests?
12. What result of the tests made him doubt his *hypothesis*?
13. Outline the *data base* that Dr. Fisher had built up, and that he used to arrive at the *hypothesis* described in the first sentence of section 5.
14. Dr. Fisher's final diagnosis was one of obstructive jaundice. Explain how this conclusion fits with the Principle of Parsimony.
15. Explain how the procedures used by Dr. Fisher fit in with the Principles of Tentativeness, Consistency, Causality, and Continuous Discovery.

5 How the Size of the Earth was Discovered

The Greek philosopher Eratosthenes was interested in almost everything. He was a historian, geographer, poet, and theatre critic. Someone like this, thought the pharaoh Ptolemy III, would be the ideal person to put in charge of the first great library in history, in the Egyptian city of Alexandria.

Eratosthenes' main interest, though was in mathematics and astronomy. He was the first person to suggest that an extra day be introduced into the calendar every fourth year to keep the calendar in line with the seasons — our leap year. In the library one day around 240 B.C., he was reading a papyrus book about Syene, a small town in the south of Egypt. One thing caught his attention in particular. At noon on midsummer day in Syene, vertical sticks cast no shadow. Also, the sun shone straight down a deep well, so that its reflection could be seen by someone peering down the well into the water.

Eratosthenes was puzzled. The book was saying that on midsummer day, at noon, the sun was shining directly downwards. If that had ever happened anywhere he'd been, he thought, he'd surely have noticed it. So next midsummer day, he set up a vertical stick in Alexandria, and watched it to see if it cast a shadow at noon. As the morning went on, the shadow became shorter. At noon, it wasn't very long, but it was still there. Eratosthenes went further. He measured the angle at the top of the stick between the end of the shadow and the stick. It was one-fiftieth of a complete circle — a little more than 7°.

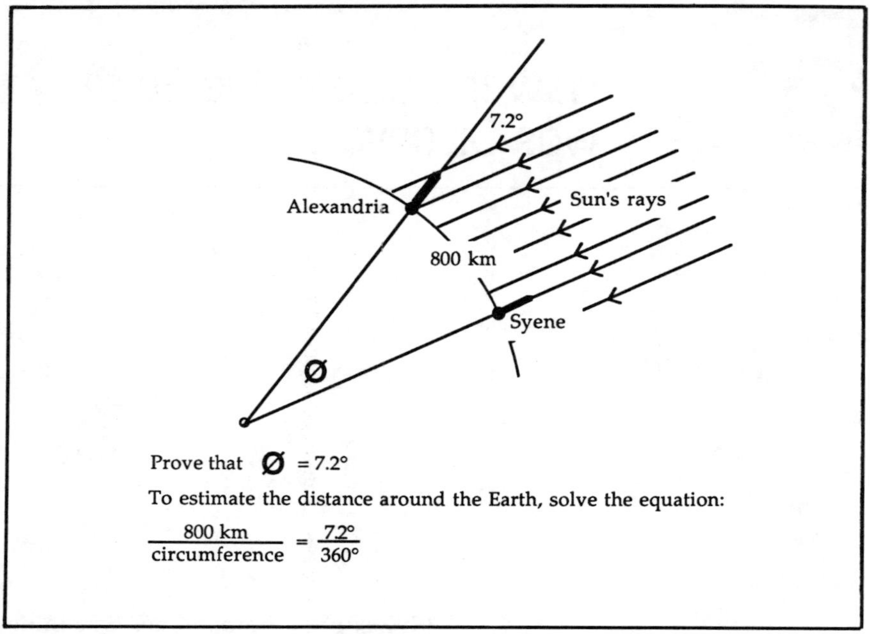

FIGURE 5-1

At noon on Midsummer's Day, the stick casts a shadow at Alexandria, but not at Syene

How could this be? Eratosthenes assumed that the sun was a very long way away, so that its rays were parallel. He could see only one explanation: the earth must be curved. This idea of curvature, he reasoned, fitted in well with something else that people had always known. If a ship had just disappeared over the horizon, it could still be seen by someone at the top of a nearby hill. If was for this reason that the lookout on a sailing ship was always sent up to the "crow's nest" at the top of the mast, high above the deck, so that he could see land as soon as possible.

Eratosthenes went one step further. Assuming that the earth is completely round — like the sun and the moon — then its curvature is the same all over. So if the distance from Alexandria to Syene represents a fiftieth of a circle (just over 7°), then the distance around the world — the circumference — must be fifty times as far. All that remained was to find the distance between Alexandria and Syene. Eratosthenes hired a man to pace it out. In our units, his estimate of the distance between the two cities would be about 800 km. Multiplied by fifty, this would make the earth about 40 000 km around — approximately the right answer. Eratosthenes speculated that all the seas on the earth were interconnected, so that it should be possible to sail right round the world. Seventeen centuries later, this feat was accomplished in an expedition led by Ferdinand Magellan.

Back in Eratosthenes' time, not everyone agreed with him. The ancients found it hard to accept that their part of the earth was such a small fraction of the whole. A century or so later, another Greek, Poseidonius, made another estimate that was more to the liking of the powers at the time. He used the position of the star Canopus in place of the sun. But he made a mistake in his calculation and came up with another figure for the earth's circumference: less than 30 000 km.

In 1492, when Columbus set sail from Portugal hoping to reach China, it was Poseidonius' calculation that he used. Columbus thought that the journey would be about 5000 km. If he'd known it was closer to four times as far, he would probably never have started out, and would never have discovered America. Magellan's voyage, beginning in 1519, finally showed that Poseidonius' estimate of the earth's size was wrong, and that Eratosthenes was right. Four of Magellan's five ships were lost in the course of the three-year voyage, and Magellan himself was killed in a skirmish with natives in the Philippines. But his expedition was the first to sail right round the world, finally proving what Eratosthenes had worked out so many years before: that the earth was round, and that the circumference was about 40 000 km.

QUESTIONS

1. Reading about the Egyptian town of Syene, Eratosthenes came across two facts which especially interested him. What were these two facts?
2. Describe what Eratosthenes did on the next midsummer day to investigate these matters further.
3. Eratosthenes' results were different from what the book said had happened in Syene. What differences did he find?
4. How did Eratosthenes explain these differences?
5. What else had to be done before Eratosthenes could calculate the size of the earth?
6. Why were Eratosthenes' conclusions not very popular in his own time?
7. If Eratosthenes' calculations had been accepted in the late 1400s, what historical event might not have taken place? Explain.
8. How was it finally shown that Eratosthenes' work was correct?

ANALYSIS

1. In the book about Syene, Eratosthenes read about two odd facts that formed part of his *data base*. What were these facts?
2. What did these facts show about the way the sun shone down on the earth?
3. What *experiment* did Eratosthenes do?
4. For this experiment, Eratosthenes must have had a *hypothesis* about the earth itself. What was this hypothesis?
5. What other well-known observation supports the *hypothesis*?
6. What *prediction* did Eratosthenes make?
7. The success of his prediction probably came as no surprise. But it was nevertheless necessary for him to do the experiment. Which of the *Ten Principles* is in operation here?
8. Sticks, wells, shadows, ships on the horizon—show how the Principle of Parsimony applies here.
9. What explanation might Eratosthenes have given for what he read in the book that would have violated the Principle of Materiality?
10. Show how the Principle of Continuous Discovery applies to this narrative.

6 Continental Drift

Many people, looking at a map of the world, are struck by the similarity in shape between the east coast of America and the west coast of Africa. The coastlines seem to fit, almost like the pieces of a jigsaw puzzle. The bulge on the eastern coast of South America, for example, seems to fit very well with the indentation on the west coast of Africa.

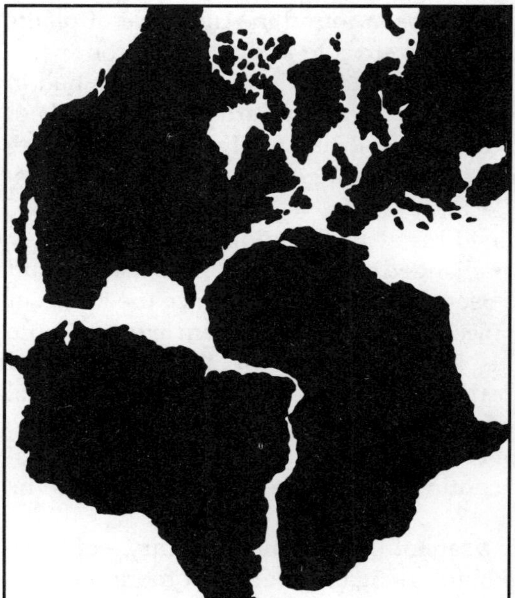

Computer reconstruction of the positions of the continents 140 million years ago.

FIGURE 6-1
Positions of the Continents

The German geologist Alfred Wegener noticed this, too. In 1915, he suggested that America and Africa might once have been joined together, and that they had drifted apart by a process he called Continental Drift. He pointed out that longitude measurements indicated that America and Europe were moving apart at a rate of about five metres a year. According to Wegener, what was happening was that the granite continents were floating apart on a ''sea'' of basalt, and that the ocean moved in to fill the gap.

Wegener's suggestion, if correct, would explain a number of things. One of these is the direction of the magnetic fields in ancient rocks. If the rocks and the poles were in the same position when these rocks were formed as they are now, then their magnetic fields would all have been directed along North-South lines. In fact, though, the magnetic fields point in various different directions in different places around the world. The rocks seem to have turned around, or else the magnetic poles were once in different positions from where they are today. Continental Drift explains this very easily: the rocks have simply drifted about on the continents.

There was also a good deal of biological evidence that the continent of America was once joined to Europe and Africa. Many similar species of animals and plants appear on both sides of the Atlantic, and it seemed sensible to suppose that they evolved from the same ancestors. Frogs, toads, and freshwater fish such as trout, pike, and whitefish are very similar in America and Europe. Many of the freshwater fish in South America are similar to other species in Africa. Since these fish die in salt water, they could not have swum across thousands of kilometres of ocean. And even in Wegener's time, fossils had been found in Antarctica of plants which could not have lived in the sub-zero cold of that barren continent.

Many scientists, though, did not accept that Continental Drift had in fact taken place. Basalt, they pointed out, is a very hard rock. It would not allow the granite continents to drift across it very easily. Enormous forces would be necessary, and Wegener had not suggested what could be causing these forces.

Wegener's opponents explained his evidence in different ways. They pointed out that the climate has changed often in the earth's history, and the Antarctic might once have been much warmer — hence the fossils of warm-weather plants. Another piece of evidence came from later and more precise longitude measurements. The earlier measurements had seemed to indicate that Europe and North America were moving apart at several metres per year. Now these were shown to be wrong. They had been one of the most important pieces of evidence which had led Wegener to propose in the first place that the continents might be moving about on the earth's surface.

However, during the middle years of the twentieth century, evidence for Continental Drift began to build up again. More and more relationships were found between animal and plant species on different continents, and a great number of fossils were found which tend to support the theory.

Fossils of the reptile *Mesosaurus*, for example, have been found both in South America and in South Africa. The giant dinosaur *Brachiosaurus* has left fossils both in North America and in Europe. Fossils of another dinosaur family, *Sauropoda*, have been discovered on every continent.

Dinosaurs are reptiles, and they just might have swum across the ocean. But the same cannot be said for amphibians, which die in salt water. The first fossilized bone of an amphibian in Antarctica was discovered in 1968. It was about 6 cm long, and belonged to an extinct species called *Labyrinthodonta*, whose fossils have also been found in South Africa. In 1982, fossilized remains of a small marsupial (a mammal which carries its young in a pouch) were found in Antarctica. The remains are identical to those of a 20 cm long mouse-like animal, *Polydolops*, which was common in South America long ago. It is difficult to escape the conclusion that Antarctica must have been joined to other continents at the time these animals lived.

Geological evidence for Continental Drift is also accumulating. The edges of the continental shelves match each other much more closely than do our present shorelines. The same type of carbonate rocks are found both in Scotland and in Newfoundland, and many rock formations on the east coast of South America match those on the west coast of Africa. Not only do the rocks match — the orientation of magnetic crystals in the rocks matches as well.

Recently, extremely sensitive instruments have detected that movement of the continents away from one another is taking place after all. The movement is much slower than had been believed in Wegener's time. The speed varies from 1 cm to 15 cm per year.

How can the continents drift apart? The University of California's deep sea drilling project has shown one possible way. The scientists found that ridges on the ocean floor are made of material that is newer than the rest of the ocean bottom. Perhaps the upthrust of this new material in the ridges is actually pushing the continents apart.

Continental Drift is now accepted by most scientists, not as fact, but as a *theory*: a hypothesis for which there is a good deal of evidence, and which makes sense of large numbers of facts. In the future, we can be sure that new evidence will be found. Perhaps this will fit in with the theory. Perhaps, though, it will not, in which case the theory will have to be altered, or even thrown out altogether. In any case, the theory will have served a useful purpose. It explains great numbers of observations in many areas of geology and biology: fossilized dinosaurs, matching rock formations, new rocks on the ocean floor. Perhaps most important, the theory also serves as both a spur and a guide to new investigations.

QUESTIONS

1. When you look at a map of the world, what observation can you make which supports the theory of Continental Drift?
2. Wegener cited another piece of evidence that Continental Drift is occurring. Later, the measurements he referred to were shown to be wrong. What was this evidence?
3. Using Continental Drift, Wegener could also explain a set of geological observations connected with the earth's magnetism. What were these observations?
4. How can Continental Drift account for the observations you described in question 3?
5. Continental Drift could also account for some biological observations connected with fish and amphibia. What were these observations?
6. How can Continental Drift account for the observations you described in question 5?
7. What geological fact caused many scientists to doubt that Continental Drift can occur?
8. Without Continental Drift, how did these scientists explain the observations you described in question 5?
9. Summarize the evidence which now leads most scientists to accept the view that Contintal Drift is occurring. Use two sub-headings: (a) geological evidence, and (b) biological evidence.
10. What evidence has been found by the University of California deep sea drilling project which might indicate *how* Continental Drift is taking place.

ANALYSIS

1. Summarize the *data base* which led Wegener to propose that the continents were moving about on the Earth's surface.
2. Later, people came to doubt that Continental Drift was in fact occurring. In terms of the *cycle of proof,* what was happening here?

3. More recently still, further evidence has been found which has made most scientists think that Continental Drift is taking place after all. In terms of the *cycle of proof,* what is happening now?

Each of the statements in questions 4 – 10 is an example of one of the *Ten Principles*. For each statement, identify the principle concerned, and explain why the statement provides an example of it.

4. Wegener was attacked by people who asserted that God had placed similar frogs, toads, and freshwater fish on many continents. He was not convinced by these arguments, and neither are most present-day scientists.
5. At first, most scientists accepted the Theory of Continental Drift. Then it fell into disrepute. Now, it is accepted again, yet scientists are not embarrassed by these about-turns.
6. Scientists don't find it difficult to believe that the continents were not always in the same positions as they are today.
7. Continental Drift explains a large number of facts, from the age of deep sea rock samples to the location of fossilized frogs.
8. Scientists believe that the continents will continue to drift apart as long as basalt continues to well up from the ocean floor.
9. Scientists are confident that new evidence concerning Continental Drift will be discovered in the future.
10. Even though there is now a great deal of evidence for Continental Drift, scientists are still prepared to modify the theory, or even abandon it, in the light of new evidence.

7 The Life and Times of Galileo Galilei

Galileo Galilei was born in 1564, the same year as William Shakespeare. He was born in northern Italy, where people were prosperous and well-educated. This prosperity[1] enabled some people, and Galileo was one, to devote their lives to purely intellectual pursuits. In less wealthy societies, almost everyone has to be engaged in more immediately practical activities such as obtaining food.

The telescope

Galileo was a mathematician, an astronomer, a physicist, and an inventor. His first published work described an ingenious balance for determining the density of precious objects, thus making it possible to find out quickly and easily what they are made from. Later, he heard about the invention of the telescope in Holland, and promptly went ahead to construct a much improved model, through which the wealthy merchants of the city could observe their ships at sea two hours before they reached port. Galileo was more interested in using his telescope to study astronomy, and it is as an astronomer that he is most remembered today. But his practical inventions contributed a good deal to his fame during his own life.

Why some things fall faster than others

Galileo's first major line of purely scientific investigation concerned the behaviour of falling bodies. Almost two thousand years earlier, the Greek philosopher Aristotle had taught that heavy objects fall faster than light ones, and this was accepted as fact in Galileo's time. The teachings of

Aristotle, in fact, had become part of the doctrine of the church, and were accepted by nearly everyone.

Galileo, though, refused to believe things just because they were written in books, or because everyone else believed them. He insisted on trying things out for himself.[2] He built an ingenious inclined track to study the acceleration of moving and falling objects. On the track, a ball would move more slowly than if it were just falling freely through the air, making it easier to study. But the ball would fall in much the same kind of way.

Galileo had at first thought that the speed of an object's fall depended upon its density. That would explain why, for example, leaves, feathers, or snowflakes fall more slowly than bricks or lead balls. But his experiments showed that wooden balls rolled down the track very nearly as quickly as lead ones, even though wood's density is much lower. Although his density idea made a lot of sense to him, he realized that he was going to have to change it.[3] The slower rate of fall for leaves, feathers, and the like now had to be explained.[4] Galileo's view was the same as ours today: the effect of air resistance. Every sailor and horseman already knew about that.[5] Galileo further concluded that in a vacuum, everything would fall at the same speed.

Galileo also noticed that balls rolled down his track at constantly accelerating speed. As long as friction was not too great, they would roll along an almost horizontal track without slowing down at all. He thus came to the conclusion that once something was moving, no force was needed to keep it going unless some other force slowed it down—such as friction or air resistance. On this point also he came into conflict with the teachings of Aristotle, and also with the established religious belief that the sun, moon, and stars had to be pushed along by the eternal labours of angels. Galileo was not satisfied with "explanations" of this kind.[6]

Galileo the astronomer

With his telescope, Galileo made a painstaking and thorough study of the heavens. He observed that there were mountains on the moon and spots on the sun. By following the sunspots from day to day, he found that the sun rotated on its axis every twenty-seven days. These observations, too, conflicted with the religious beliefs of the time. The church taught that the heavens were perfectly orderly, and that only on earth did irregularity and disorder occur. Again, Galileo preferred his own observations to the established belief.[7] But already, he was making enemies among those who preferred the established view, and these enemies were beginning to store evidence against him.

Another of Galileo's observations also gave ammunition to his enemies, even though it was something that everyone could see. It happened during 1604 that a new star (Kepler's Nova) appeared. For a while, it was by far the brightest object in the night sky. The appearance of this object was determinedly ignored by those who thought that the heavens were perfectly and eternally unchanging. But the appearance of something new did not bother Galileo at all.[8] It fitted in quite well with his view that some rules of nature were followed in the heavens just as they were here on earth.[9] Since these laws gave rise to variable and irregular occurrences on earth, he reasoned, it was not surprising that the same thing happened in the faraway regions of space.

With his telescope, Galileo was able to observe many new features of the night sky. Through the telescope, the stars all remained mere dots of light, although many more of them could be seen. But the planets appeared as small disks or globes. Galileo concluded that they must be much closer to the earth than the stars are.

The moons of Jupiter

Two further discoveries were among Galileo's greatest contributions to astronomy, but brought him into even greater conflict with the established ideas of the day. The first was his discovery of the moons of Jupiter; the second concerned the planet Venus.

Galileo's telescope revealed to him that the planet Jupiter had four moons revolving around it. This was definite proof that not all bodies in the sky circled around the earth, as the established theory held.

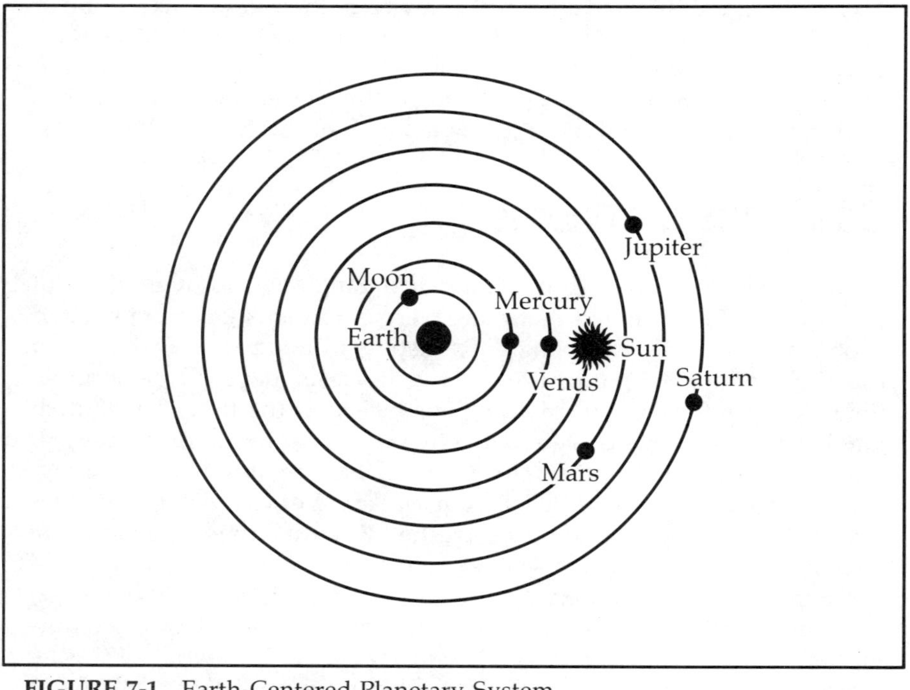

FIGURE 7-1 Earth-Centered Planetary System

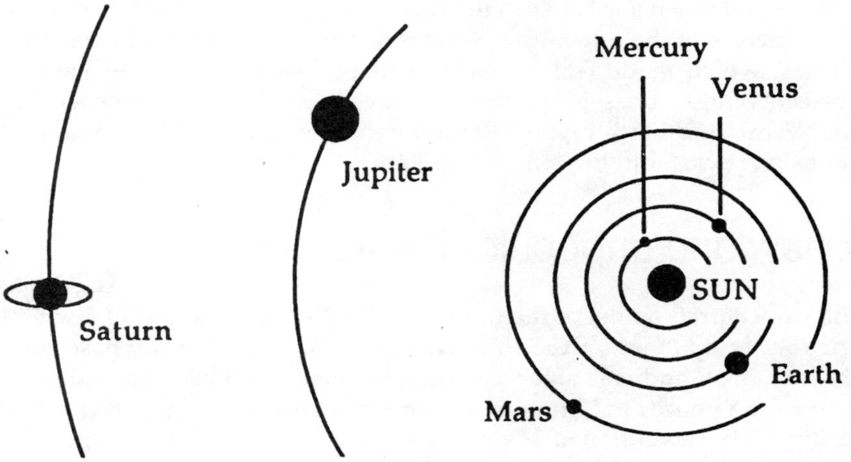

Copernican view of the planetary system, showing the planets known in Copernicus's time. The sizes of the Sun and the planets are not drawn to scale with the planetary orbits.

FIGURE 7-2 Sun-Centered Planetary System

Half a century earlier, the Polish astronomer Copernicus had put forward the idea that the sun, not the earth, occupies the centre of the universe. Galileo's new observations of Jupiter's moons fitted this theory. He reasoned that if our moon can revolve around the earth, so can Jupiter's moons revolve around Jupiter. In both cases, a relatively small body revolves around a much larger one. The relatively small moon revolves around the much larger earth. Similarly with Jupiter's moons: they are smaller than Jupiter, and revolve around it. How natural, then, to assume that the sun and the planets behave in a similar way,[10] with the enormous sun at the centre of the circle around which the planets move. This would include the earth, of course. Another observation added support to the sun-centred theory of the solar system. Galileo found that the moons further from Jupiter take longer to go around it than do the ones closer in. This observation corresponded well with Copernicus's calculations, which showed that planets far from the sun must take longer to make their orbits around it—if indeed the sun really is at the centre of the solar system.

Venus

The second of Galileo's great discoveries about the planets concerned Venus. It appeared much larger at some points in its orbit than at others. This is just what would be expected if both Venus and Earth revolved around the sun at different rates. At times, Venus would be closer to the earth than at other times, and would then look larger. Also, Venus shows crescent phases at certain times during its orbit, just as the moon shows as a crescent when it is between the sun and ourselves. Under the conventional picture of the solar system, there was no reason for these crescent phases, which made Galileo suspicious of this earth-centred view.[11] Crescent-shaped phases, however, are exactly what would be expected if both Venus and Earth revolved at different rates around the sun, and if Venus were closer to the sun.

A new and unpopular theory

The sun-centred model of the universe, then, passed the test of a scientific theory by meeting the following criteria. It was firmly based on observations[12] and was able to account for many of these observations, such as the phases of Venus and its apparently varying size. It did not require the intervention of any supernatural beings,[13] assuming that everything that happened did so as a result of the operation of natural laws.[14] The theory explains a large number of observations with a few simple ideas,[15] but rejects the dogmatic and absolute assertions of the old earth-centred theory. For Galileo, nothing was absolute. To understand the behaviour of Venus, for example, the motions of both Venus and Earth had to be taken into account. Predicting the positions of Jupiter's moons needed a knowledge of how they moved around Jupiter and of how Jupiter moved around the sun. And since the observer is on earth, which itself is revolving around the sun and rotating at the same time, the situation is quite complicated, even though the basic ideas are very simple. In the sun-centred theory, everything seemed to depend on something else.[16] This, perhaps, was the aspect of the theory which most disturbed church authorities.

The new theory, put forward by Copernicus and further explored by Galileo, was a revolution in thinking. For thousands of years, everyone had regarded the earth as being the centre of everything. The new theory set the stage for a multitude of new discoveries,[17] particularly those of Johannes Kepler (with whom Galileo corresponded for a while) and perhaps the greatest scientist of all time, Isaac Newton. An account of the work of these astronomers appears later in this book.

Galileo was often scornful in his rejection of established doctrines. He was impatient with people who could not (or would not) see his reasoning. He made many enemies among the ranks of powerful church dignitaries. In 1616, the sun-centred theory was forbidden by the church. In particular, Galileo was forbidden by the Inquisition to explain it, to defend it, or even to think about it. For several years, he obeyed the first two parts of this order. He bided his time until things became a little less rigid in the church. In 1623, it seemed that this was happening. Maffeo Barberini, whom Galileo knew and respected, became pope. In the following year, Galileo came to Rome, and had a long series of conversations with the new pope. It turned out that he had been too optimistic. Not only did the pope stand behind the original prohibition; he also ordered Galileo to include statements in his forthcoming book which would assert that the universe was operated by God on a detailed day-to-day basis in a way that humans could never understand or explain — exactly the opposite of what Galileo believed.

Nevertheless, Galileo persisted in his optimism that the pope was too intelligent not to see his reasoning, and that the pope's attitude was merely the result of the conservative political pressure of Galileo's enemies. Galileo continued work on his major book on astronomy, and in 1632 he felt confident enough to publish it. He had managed to get it approved by no fewer than four church censors.

Galileo in trouble

The book created a great storm. Entitled *Dialogue on The Two Chief World Systems*, it portrayed one character arguing for the old earth-centred view of the universe, and the other arguing for Copernicus's sun-centred view. Everything the pope had ordered Galileo to include in the book was put into the mouth of the character, called Simplicius, who argued for the old system. These arguments were far weaker than those of his opponent, though Galileo did have Simplicius unconvincingly win the argument in the end. Galileo's trick angered the pope. He may have suspected that Simplicius was intended as a caricature of himself. Galileo was brought before the Inquisition in 1633 and threatened with physical torture. He was seventy, and he had the example of the scientist Giordano Bruno to convince him that the Inquisition meant business. Bruno had been burned alive at the stake in 1600 because of his opinions.

Publicly, Galileo reversed his views. We cannot tell what his private views were, but legend has it that when he rose from his knees at the end of his statement to the Inquisition, he muttered beneath his breath, ''Eppur si muove'' (it still does move), referring to the earth.

Galileo was sentenced to confinement in his house for the rest of his life. He continued working, but had a difficult time getting his work published. His last book, on forces and motion, was finally published in 1638 by Protestants in Holland, two years after it was finished. Galileo, unhappy and by now totally blind from his observations of the sun years before, died in 1642.

The effect of his trial was to bring to a halt all important scientific work in southern Europe.[18] From then on, the focus of scientific endeavour moved to Holland, Germany, and Britain. But Galileo's discoveries were the foundations upon which modern astronomy and modern mechanics were constructed. His greatest influence, perhaps, lay in his belief in the reliability[19] and universal application of natural laws, and his insistance upon careful observation[20] and logical reasoning to find out about these laws — what we have called the *cycle of proof*.

Of course, the Catholic Church today does not behave in the way it did in the 1600s. Galileo's *Dialogue* was removed from the list of books prohibited to Catholics in 1835. In 1965, on a visit to Pisa, Pope Paul VI went out of his way to speak highly of Galileo, thereby making clear that the church now no longer feels the conflict it once did between religious beliefs and the methods and conclusions of science. In 1979, the church appointed a commission to study the work of Galileo and the church's attitude towards it. Finally in 1983, in an address to a large group of prominent scientists, Pope John Paul II clarified the church's latest view. ''We cast our minds back to an age where there had developed between science and faith grave incomprehension,'' he said, ''which only humble and patient re-examination gradually succeeded in dispelling. . . . The church herself learns by experience and reflection, and she now understands better the meaning that must be given to freedom of research.''

Even after 350 years, Galileo's work still commands attention and respect from the institution which was once his most powerful opponent.

QUESTIONS

1. Name two inventions made by Galileo, and explain why they were useful in Galileo's time.
2. What had Aristotle taught about the behaviour of falling bodies?
3. Describe what Galileo discovered about the behaviour of these falling bodies and give an account of the experiments which led to these discoveries.
4. Give an account of some of Galileo's astronomical observations (other than those on the moons of Jupiter or on Venus) and explain how these observations did not fit in with the accepted beliefs of the time.
5. Describe what Galileo observed about the behaviour of the moons of Jupiter and explain how these observations did not fit in with the accepted beliefs of the time.
6. Describe two observations made by Galileo of the planet Venus, and explain how these observations fit in better with the sun-centred theory of the solar system than they do with the earth-centred theory.
7. What orders were given by the church to Galileo in 1616?
8. What trick did Galileo use in his 1632 book to get around these orders?

ANALYSIS

Each numbered superscript (1 – 20) in the article points out an example of one of the *Ten Principles* in operation. For each number, identify the principle which applies, and explain why.

8 History of Modern Theories About Heat

Today the difference between heat and temperature is well understood. We define temperature as the degree of hotness of something, and we know for example that if a solid object is at a high temperature, its molecules or ions are vibrating more quickly than when it is at a low temperature. Heat, on the other hand, is a form of energy. An object's temperature is only one of the factors which determine how much heat it holds. Another is the mass of the object: a heavy object holds more heat energy than a light one. The third and last factor is the material the object is made of: different materials can hold different amounts of heat. Water holds most, and metals least. The ability of a substance to hold heat is measured by a property known as the specific heat capacity of the substance. Water has a specific heat capacity of 4200 J/(kg.°C), for example, while the specific heat capacity for lead is only 130 J/(kg.°C). A kilogram of water holds over thirty times as much heat as a kilogram of lead at the same temperature.

We know too that other forms of energy can easily be converted to heat energy, and that with rather more difficulty heat energy can be transformed into other kinds of energy, such as kinetic energy (energy of motion) or electrical energy.

People haven't always known all this. Like many of our fundamental scientific concepts, these ideas seem simple to us now, because someone has explained them to us. There was nobody to explain them to the original researchers.

The first person to realize that different substances have different specific heat capacities was the Scotsman, Joseph Black (1728 - 1799). Black performed experiments, similar to those often done in high school laboratories now, which showed these differences. Black showed that raising the temperature of iron needed more heat than for the same mass of lead. Similarly, he showed that more heat was required to raise the temperature of a certain mass of iron by 1° than for the same mass of lead. Black called this property of substances "capacity for heat." It is the same property which we now call specific heat capacity.

Black was also the first person to see the difference between temperature, which he called "intensity of heat," and heat itself. He regarded heat as a substance: an "imponderable fluid." The French chemist Antoine Lavoisier (1743–1794) saw it in the same way; he called heat "calorique," and went so far as to include it in his list of the chemical elements, along with lead, oxygen, and about thirty others.

This was the Caloric Theory of heat. A hot object, according to the theory, is one which contains a great deal of caloric, whereas a cold one contains very little. Heating an object up means adding caloric to it; cooling it down means allowing the caloric to escape from it.

The Caloric Theory accounts rather well for a rather large number of observations and experimental results. Predictions made from the theory about what happens when hot and cold water are mixed, for example, turn out to give the right answer for the final temperature. Using the Caloric Theory, the same set of specific heat capacities which we use now can be determined by the same experiments we do now. The theory made sense.

Benjamin Thompson (1753–1814) was an American, but one who supported the Loyalist (British) side during the Revolution. The result was that he was forced to spend most of his life in exile in Europe. For several years after 1780, he worked for the Elector Karl Theodor of Bavaria, who made him a count in 1790. He chose the name Rumford, after his birthplace Rumford (now Concord), New Hampshire.

In 1798, Rumford was supervising the boring of brass cannons, and he noticed that great amounts of heat were produced. He had to pour pails of cold water over each cannon to prevent it from becoming red hot. The Caloric Theory's explanation for this was that the boring process "loosened" the caloric as the metal was broken down into shavings. But Rumford noticed that there seemed to be no limit to the heat that could be produced when the boring continued for a long time. In fact, when he tried boring with an instrument so blunt that no shavings were produced, the metal became even hotter—contrary to the predictions of the Caloric Theory.

Rumford could not see where all the caloric could be coming from. He decided that heat wasn't a substance at all. Rather, it was a form of vibratory motion. When the brass was hot, the particles it was made from were vibrating quickly. When it was cold, they vibrated more slowly. When the borer rubbed against the cannon, friction caused the particles to vibrate faster, and the cannon became hot.

The following year, in England, Rumford tried more experiments. He weighed a block of ice, and then melted it. According to the Caloric Theory, the mass of the ice should increase, since caloric had been added. But the mass turned out to be the same as before. Thus caloric, if it existed, must be a substance without mass—an unlikely notion.

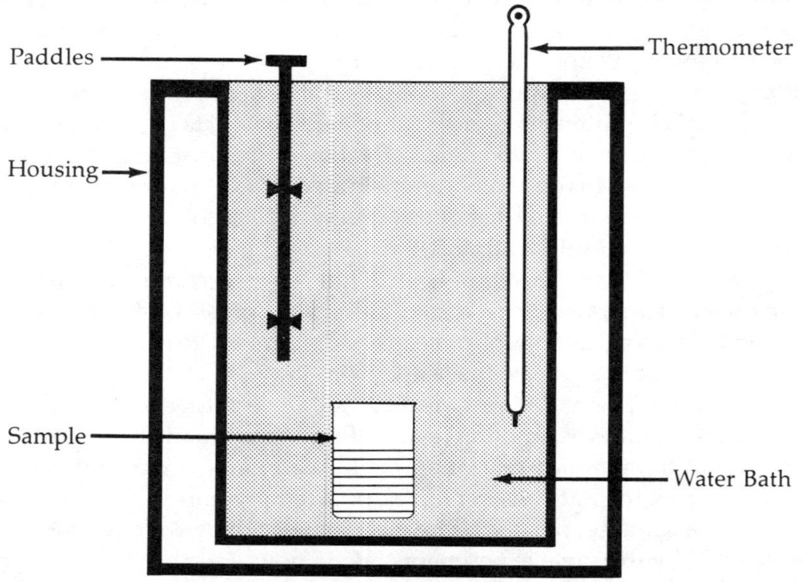

A calorimeter is used to measure the amount of heat energy in known amount of a substance by distributing the heat through a fluid and measuring the rise in fluid temperature.

FIGURE 8-1 Principle of the Calorimeter

A year or so later, the English chemist Humphrey Davy (1778 - 1829) performed an even more conclusive experiment. According to the Caloric Theory, ice below 0°C would contain very little caloric — certainly not enough to melt the ice. Davy arranged a mechanical device to rub two pieces of ice together at a temperature just below 0°C. Enough heat was produced, he found, to melt some of the ice. Again, this result did not fit well with the Caloric Theory, and Davy came to the same conclusion as Runford: that heat was a form of motion, and not a substance at all.

Over half a century later, another Englishman, James Joule (1818 - 1889), rounded out the picture by making it quantitative — attaching numbers to it. Joule heated water by churning it vigorously with paddles; he measured the amount of work being done on the water by the paddles and the amount of heat energy produced in the water. He squeezed water through small holes: energy was necessary to do this, and this energy heated up the water. And he measured (on his honeymoon!) the temperature at the top of a waterfall and at the bottom; the kinetic energy of the falling water, he reasoned, should be converted to heat energy when its motion was stopped at the bottom of its fall. By a variety of means such as these, Joule was able to show exactly how much mechanical work was equivalent to a certain amount of heat energy.

In the latter part of the nineteenth century, the Scottish mathematician James Clerk Maxwell (1831 - 1879) and the Austrian physicist Ludwig Boltzmann (1844 - 1906) worked together on a sophisticated mathematical analysis of the exact energies and speeds of each particle in any substance at any temperature.

The Maxwell-Boltzmann Kinetic Molecular Theory has been most useful to both physicists and chemists ever since. It helps us understand the speeds of chemical reactions, how increasing the temperature makes these reactions go faster, and the action of catalysts — substances which speed up chemical reactions without being consumed themselves. The theory, though, rests on the fundamental conclusions of Rumford, Davy, and Joule: that heat is a form of energy, and temperature is a measure of the kinetic energy of the individual particles — atoms, molecules, or ions — which make up all matter.

QUESTIONS

1. What is the difference in behaviour between the molecules in a hot object and the molecules in a cold one?
2. Apart from temperature, what two factors go into determining the amount of heat stored in a hot object?
3. The difference between heat and temperature is easy for us to understand. Why has it not always been obvious to people?
4. What two discoveries about heat were made by Joseph Black?
5. According to the Caloric Theory, what is the difference between a hot object and a cold one?
6. Describe an experiment for which the Caloric Theory gives the correct results.
7. Describe the experiments performed by Count Rumford, for which the Caloric Theory gives no easy explanation.
8. Describe an experiment performed by Humphrey Davy, for which the Caloric Theory also gives no easy explanation.
9. Give an account in your own words of the contribution to the theory of heat made by James Joule.
10. Give an account in your own words of the contribution to the theory of heat by James Clerk Maxwell and Ludwig Boltzmann.

ANALYSIS

Questions 1 – 6 refer to the Caloric Theory. State which part of the *cycle of proof* is represented by each of the six statements, and explain.

1. Joseph Black used hot water to heat up samples of various metals, and found out how much each of the metals rose in temperature.
2. Black thought of heat as a fluid. A hot object would contain a lot of it, a cold object not so much.
3. Rumford observed closely while brass cannons were being bored.
4. There seemed to be no limit to the heat that was produced — especially when a blunt drill was used.

5. According to the Caloric Theory, a block of ice should get heavier when it is melted.

6. Rumford found that blocks of ice had the same mass after they had melted.

7. Black and Lavoisier, who shared his belief in the Caloric Theory, are now thought to be wrong. Does this mean that they made no contribution to our modern understanding of heat? Explain.

8. Analyze the Kinetic Molecular Theory of Heat in terms of the *cycle of proof*. Do this by stating the *hypothesis*, and some of the facts in the *data base* used to arrive at it. Describe an *experiment* which Humphrey Davy did to test the hypothesis, and the *prediction* which Davy made about the result of this experiment. Explain what this result does to our confidence in the hypothesis.

Each of the statements 9 – 12 is an example of one of the *Ten Principles*. For each, state which of the principles is referred to, and explain.

9. Since the time of Rumford, Davy, and Joule, a very large number of observations about heat have been made. The Kinetic Molecular Theory, which basically is fairly simple, is able to explain all of them.

10. Everyone working in this field expects the specific heat capacities of various substances to be the same now as they were in Rumford and Davy's time.

11. After Rumford and Davy came Joule, Maxwell, and Boltzmann, who extended the Theory and made it quantitative. Scientists expect theories to grow in this kind of way.

12. Even now, with the Kinetic Molecular Theory so well established, scientists would be prepared to modify it or even discard it, if experiments yielded results which didn't fit it.

9 Insulin: The Work of Frederick Banting

Born in Alliston, North of Toronto, Ontario, Frederick Banting obtained his medical degree from the University of Toronto in 1916. After serving as a medical officer in the First World War, Banting went to London, Ontario, and practised as a surgeon. One of his interests was diabetes, which causes people to be abnormally thirsty and to feel more and more tired as the disease progresses. At the time, there was nothing to help diabetics. Patients slowly but surely became thinner and thinner, and eventually died.

In diabetics, the amount of glucose in the blood reaches unusually high levels. Eventually glucose appears in the urine. Clearly the diabetic is wasting glucose, and is therefore not getting the full benefit from the carbohydrate part of his or her diet.

That much was known in Banting's time. And it was suspected that the pancreas had something to do with diabetes. The pancreas, it was known, is a digestive organ whose main function is to produce digestive juices to break down proteins into amino acids, which can then be absorbed in the intestine. It had been discovered that when the pancreas was removed from dogs, diabetes developed. Physiologists wondered, therefore, if the pancreas had another function in addition to producing digestive juices. It might, they thought, be producing a hormone to control the breakdown of glucose molecules. This breakdown is a central part of carbohydrate metabolism, which extracts the energy we need from our food. If not enough of the hormone was produced, the glucose would stay in the bloodstream, and could not be used by the body for energy. The supposed hormone had been given a name: insulin.

The problem was that no evidence could be found for the production of such a hormone. The only chemicals to come out of the pancreas seemed to be the digestive juices. Another puzzle was that inside the pancreas

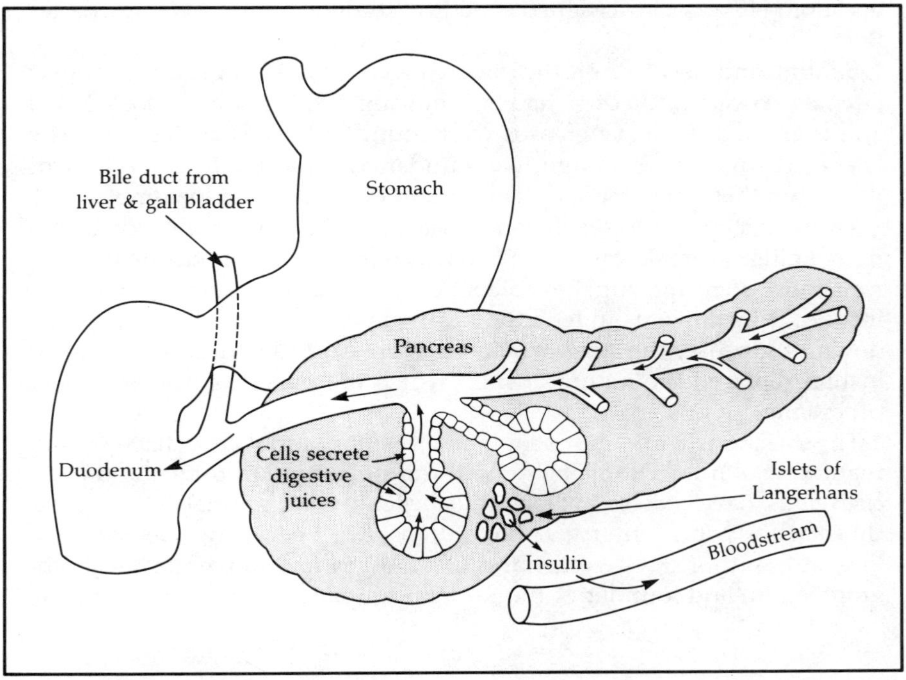

FIGURE 9-1 Location of the Pancreas

were a number of areas, different from the rest, which did not seem to be connected with the production of digestive juices at all. These areas were named Islets of Langerhans, after the German physiologist who discovered them.

It could be that these areas produced insulin, but that when people tried to extract it, it was destroyed by the digestive juices from the rest of the pancreas. So ran the thinking at the time.

In 1920, Banting read an article describing how tying off the tube leading from the pancreas to the intestine caused the pancreas to shrivel up. There was nothing in the article to say what happened to the Islets of Langerhans, and this gave Banting the idea for an experiment. If the pancreas was tied off, the main part would shrivel up and the digestive juices would no longer be produced. The Islets of Langerhans, though, have nothing to do with the digestive juices. Therefore, reasoned Banting, they should remain undamaged. And if their function is to produce insulin, they should carry on doing it. But now there would no longer be any digestive juices to destroy the insulin. Thus it should be possible to extract insulin, and perhaps to use it to cure diabetic patients.

That was Banting's reasoning, and in 1921, he went to the University of Toronto with his idea. With some difficulty, since he was no longer connected with the university, he managed to persuade Professor John Macleod to lend him a laboratory for the summer, while the professor was on vacation. He was also assigned a medical student as an assistant: Charles Best.

Banting and Best tied off the pancreases of several dogs. Seven weeks later, as expected, the dogs had become quite ill. When they were killed and examined, their pancreases were found to have shrivelled up. The Islets of Langerhans, though, were undamaged, as Banting had hoped. And from these pancreases, Banting and Best extracted insulin.

The next step was to see if insulin was indeed the hormone which controlled glucose breakdown. The two researchers removed the pancreases from more dogs, making them diabetic. Then the dogs were injected with insulin. If insulin was indeed the hormone which controlled the breakdown of glucose, the dogs would recover. And they did. The injected insulin replaced the natural insulin which the pancreas was no longer producing.

The injections work with people too. Insulin cannot cure diabetes, but regular insulin injections, along with careful attention to diet and exercise habits, have enabled millions of diabetics to enjoy normal lives. Among those who would otherwise have died have been King George V of England, author H.G. Wells, and George Eastman, inventor of photographic film and founder of the Kodak Company.

QUESTIONS

1. In what ways does a person suffer who has diabetes?
2. When a person has diabetes, what changes can be seen in the composition of the blood?
3. Explain the function of the main part of the pancreas.
4. Why did physiologists in Banting's time suspect that the pancreas had some other function besides the one you described in question 3?
5. What name was given to the parts of the pancreas whose function was not known at that time?
6. Banting and Best discovered what the function of these parts is. What is this function?
7. When a diabetic is injected with insulin, this insulin replaces the insulin which most people produce naturally. What does insulin do?
8. Apart from regular injections of insulin, what two other precautions must a diabetic person take?

During Banting and Best's experiments, a number of dogs suffered great discomfort, and eventually many of them died. Experiments like this are still done today. Answer *either* question 9 *or* question 10.

9. Some people take the view that experiments in which animals suffer pain and danger should be prohibited by legislation. Write a letter to the local newspaper supporting this view.
10. Most researchers in physiology and in medicine hope that this type of experiment will continue to be permitted. Write a letter to the local newspaper supporting this view.

ANALYSIS

Questions 1–5 concern the first *experiment* performed by Banting and Best.

1. What was the *hypothesis* being tested in this experiment?
2. State some of the facts from the *data base*.
3. Describe the *experiment* itself.

4. What *prediction* had Banting made about the experiment's results?

5. What were the results of the experiment, in terms of Banting's *confidence* in the *hypothesis*?

6. Describe Banting and Best's second experiment, using the same headings as in questions 1 – 5.

Each of the statements in questions 7 – 10 contradicts one of the *Ten Principles*. Since doctors share many of the attitudes of their scientific colleagues in the laboratory, these statements would be surprising for a doctor to make. For each statement, state which of the *Ten Principles* is being violated, and explain.

7. He has this disease because it's God's will that he should get it.

8. We'll hurry up his death a bit, so we can have a really good look at his pancreas.

9. There's no point in giving him any tests; I'm certain it's diabetes.

10. There's no reason why he got the disease. Some people get it; others don't.

10 The Discovery of Penicillin

In 1928, the Scottish biologist Alexander Fleming was performing some fairly routine investigations on a type of bacterium called staphylococcus. "Staph," as they are often called, are the type of bacteria which destroy white blood cells. The remains of these white blood cells form the pus which appears in infected wounds. When placed in a nutritive jelly or agar, these "staph" bacteria form cultures a few millimetres in diameter, each containing thousands of individual bacteria.

Fleming had finished with some dishes containing these cultures, but some of the dishes had become contaminated with mould. When he came to clean the dishes, he noticed something that surprised him. There were a number of small clear patches where the bacteria seemed to have disappeared. Many people would have ignored this observation, or would have dismissed it as being of no account, but Fleming was curious.

He investigated the clear spots and discovered that at the centre of each one was some bread mould. It seemed that the bread mould had prevented the bacteria from spreading.

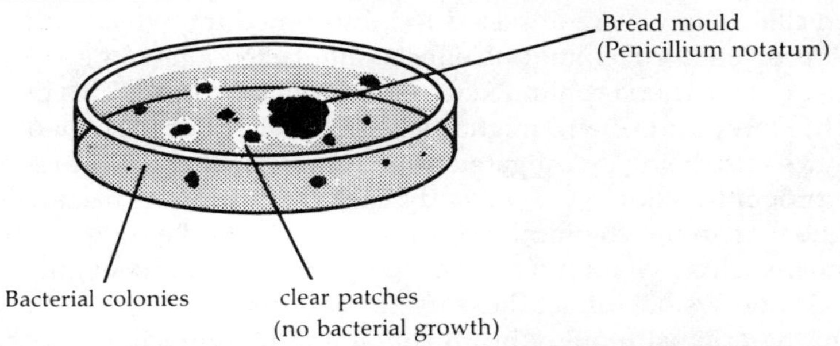

FIGURE 10-1 Discovery of Penicillin

There are many different kinds of bread mould. Fleming performed further investigations to find out which one had stopped the spread of his bacteria. He finally found that it was a strain called *Penicillium notatum*. He concluded that what killed the bacteria must be a chemical that the mould produced —something in its waste products. He prepared an extract of what he thought would be the active substance produced by the mould: the substance that was affecting the bacteria. This substance he named penicillin. He tested its effect on several types of bacteria, and found that many would not grow where penicillin was present.

Even though penicillin seemed effective against many types of bacteria, it would not be much use in medicine if it also destroyed human cells. When he checked to see whether the chemical affected white blood cells, he found that it didn't.

Fleming was not a chemist, and although he published his findings later in 1928, no one was interested enough to purify penicillin and to find out exactly what it was. Fleming himself did use it in his laboratory work. He prepared a pure culture of influenza by killing off everything else in the culture with penicillin. But as far as medical uses were concerned, a belief widely held by the medical researchers of the day made them not very interested in it. Any chemical powerful enough to kill bacteria, they reasoned, must also be fatal to the humans or other animals which had the bacteria living in their bodies.

A decade or so later, however, two developments changed the picture. First, chemicals called sulfa drugs had been shown to kill another type of bacterium, streptococci, without harm to the infected animal or person. The success of these drugs changed the way physicians and researchers regarded substances which were effective against bacteria. Clearly, contrary to the previous view, it was possible that some chemicals could be used against bacteria without doing just as much damage to the patient as they did to the bacteria. Also, the coming of World War II in 1939 made it seem likely that large numbers of soldiers would be wounded. In World War I, infection had killed many injured soldiers, and anything that could reduce the number who died would be of great benefit.

The English biochemist Howard Florey and his colleague Ernst Chain (a German who had come to England in 1933 to escape the Nazis) worked on penicillin. They were attracted to it by something which had earlier been a problem for Fleming: its effects only lasted a few hours or days. After that, the bacteria continued to reproduce and spread. This property, thought Florey and Chain, might make penicillin safer than many other substances also being investigated. Many bacteria, after all, are necessary to the proper functioning of the body itself. For example, bacteria in the intestine help in the chemical breakdown of food and also help produce vitamins, such as vitamin K. Drugs to combat infections might, if they stayed in the system, affect these useful bacteria too.

From the original mouldy broth containing the penicillin, Florey and Chain were able to obtain a pure sample. A yellow powder, it had exactly the properties which the biochemists had hoped for. Without doing much harm to whomever took it, it was effective against many different kinds of bacteria, including those causing pneumonia, syphilis, scarlet fever, and meningitis.

Scarlet fever and meningitis have been virtually eliminated by penicillin and the other antibiotics, such as terramycin and aureomycin, whose discovery rather quickly followed. Penicillin seemed almost miraculous to the doctors and veterinarians who first used it. Hardly any diseases seemed immune. But, over a period of years, many of the first antibiotics became less effective. The antibiotics themselves were just as good, but the situations in which they were used gradually changed. Strains of bacteria evolved which were resistant to the original drugs. Penicillin was so widely used against a great variety of diseases that some bacteria developed a resistance to it. To deal with these "superbugs," new antibiotics had to be developed. The process is continuing. New and tougher bacteria keep evolving, and they need new antibiotics.

Although there are now many other antibiotics, penicillin is still one of the most effective drugs in the fight against many bacterial diseases. Many people are walking around today who owe their lives to it.

QUESTIONS

1. Explain in one or two sentences the significance of each of the following in your study of penicillin:
   a) Staphylococci;
   b) Pus;
   c) *Penicillium notatum;*
   d) How *penicillium notatum* kills bacteria.
2. What commonly-held view discouraged medical researchers from investigating penicillin in the 1930s?
3. What two circumstances led to the further work on penicillin, performed by Florey and Chain?
4. The effects of penicillin do not last more than a few days. Why is this a good thing when the drug is used?
5. Why is penicillin less effective now than it was in 1940?

ANALYSIS

Each of the statements in questions 1–6 provides an example of one of the *Ten Principles*. For each, identify the principle concerned, and explain why the statement is an example of it.

1. It did not occur to Fleming that evil spirits might have killed the bacteria.
2. He was confident nevertheless that there was some reason for the clear patches in his bacterial cultures.
3. He wanted to find out more, so he performed experiments to see exactly what type of bread mould caused the clear spots.
4. In 1939, it seemed likely that war would break out. It seemed important to find ways of killing bacteria that infected wounds.
5. The success of penicillin led to the development of many other antibiotics.
6. Bacteria have changed since Florey and Chain's time. They are now more resistant to antibiotics.

Questions 7 – 10 refer to Fleming's original discovery of penicillin.

7. What was Fleming's *hypothesis* about what was causing the clear patches in his bacterial culture?

8. What were some of the facts from Fleming's *data base* which he used in arriving at his hypothesis?

9. What *experiment* did he perform to test his hypothesis and to find out more?

10. By 1939, researchers' views on possible uses for penicillin had changed from what they had been ten years before. Explain what the change was, why it occurred, and how it provides an example of the Principle of Tentativeness.

11 Mosquitoes and Repellents

There are about 2000 species of mosquitoes in the world. They feed on plant juices, nectar, and blood, sucking the liquids up through the proboscis, a long tube which sticks out from the front of the head and which contains the mouth parts. The proboscis is hard enough and strong enough to pierce the skin and blood vessels of animals, including humans. The female needs a meal of blood in order to mature her eggs, and this is why mosquito bites are so common in spring and summer, wherever mosquitoes live and reproduce. In some places, mosquitoes are more than annoyances. They can transmit diseases like yellow fever and malaria from one person to another.

Most people assume that mosquitoes find people to bite by being sensitive to their smell. They notice, for example, that mosquitoes bother some people more than others, and that hot sweating skin seems particularly attractive to them. If this is so, then insect repellent might work by masking human odour so that mosquitoes can no longer smell us. Alternatively, the repellent's smell might be unpleasant to mosquitoes, causing them to fly away. There have been many attempts to find odours which mosquitoes are sensitive to—odours both of animals and of repellents. All these attempts, though, have failed. There is no evidence that mosquitoes have a sense of smell.

Carbon dioxide

Alternatively, the carbon dioxide which warm-blooded animals breathe out might have something to do with attracting mosquitoes. In the early 1970s, a team from the British Columbia Research Council in Canada set out to study this question. The team, led by Philip N. Daykin, placed about a hundred mosquitoes in a cage and observed their activity. It took about an hour, they found, for half the mosquitoes to leave their original position and fly to another part of the cage.

If carbon dioxide had anything to do with attracting mosquitoes, then raising the carbon dioxide level in the cage would make the mosquitoes more active. Daykin and his fellow researchers tried doing this, and it worked—at least for a time. Instead of taking an hour for half the mosquitoes to move, it now took only about five minutes. But when the mosquitoes were left in this carbon dioxide-rich atmosphere for a while, their activity gradually decreased, until again it took an hour for half of them to move.

Perhaps, then, it was the increase in the carbon dioxide level which produced the added activity, rather than the actual amount. To find out, the researchers ran further tests with even more carbon dioxide. Indeed, the mosquitoes became more active at first, but then settled down to their original level of activity. They were *adapting* to each new level of carbon dioxide, in the same way as humans can adapt to a smell, a bright light, or a noise. For example, we may notice the noise of a refrigerator when we first go into the kitchen. Then we tend to forget about it, and only notice it again if the noise stops or changes.

Warmth and moisture

Another way for mosquitoes to be alerted to the presence of a warm-blooded animal might be the warmth and moisture it gives off. Daykin's colleague, R.H. Wright, devised a way of finding out if mosquitoes react to warm, moist objects. He placed three small cylindrical targets in a wind tunnel containing large numbers of mosquitoes. One target was warm, another wet, and a third both warm and wet. He activated the mosquitoes by raising the level of carbon dioxide, and they flew towards the cylinders. The results were dramatic, as the following table shows:

	warm target	wet target	warm & wet target
number of mosquitoes landing	7	22	358

FIGURE 11-1

Further experiments demonstrated that the warm and wet target was at least as attractive as a human hand. In Wright's words, there was no need to postulate a special skin odour as an attractant, and no need to look for one."

Other experiments performed at about the same time showed that some people attract more mosquitoes than others. If Wright's idea was correct, then people who attracted mosquitoes more should turn out to have warmer and damper skin than the average. And indeed, this turned out to be so.

How do mosquitoes find you?

How do mosquitoes find the warm and wet target? Perhaps they use the convection currents given off by objects which are warmer or damper than their surroundings. (Warm moist air rises because it has a lower density than dry cool air. You can see the convection currents just above the surface of a hot road in summer or above a warm drink.) Wright decided to find out how mosquitoes respond to warm moist convection currents. He describes the experiment in the following way:

> *We released a stream of warmed and humidified air from a slit in the wind tunnel, together with a trace of nonrepellent smoke to show us where the warm and wet air was going. We then photographed the flight path of a mosquito. . . .*
>
> *The flight of a mosquito initially is at random. When the mosquito encounters a wet and warm convection current, it moves steadily forward. If it passes out of the convection current into cooler and drier air, it turns. The turn is not always in the right direction or sharp enough to bring the mosquito back into the convection current, but it works often enough to enable most mosquitoes to find their prey most of the time.*

Wright summarizes the findings of these experiments, which actually took several years of work, like this:

> *A rise in the level of carbon dioxide alerts the mosquito to the approach of a suitable host. The mosquito starts flying, and when it encounters a warm and wet convection current, it keeps on going. When it begins to move out of the convection current, it turns, and the turn usually keeps the mosquito within the current. It follows the current to its source and lands.*

How repellents work

Previous research at the British Columbia Research Council had demonstrated that mosquito repellents can neutralize the effect of carbon dioxide, so that fewer mosquitoes are activated as the animal or person walks by. But as every watcher of television commercials knows, repellents do more than this. Even mosquitoes already in flight are much less likely to land on the skin if a repellent is being used. Researcher T.E. Kellogg, also of the British Columbia Research Council, decided to use modern microelectronics to help find out how this happens.

Kellogg used electrodes small enough to go into the base of one of the sensory hairs of the mosquito antenna. The first task was to find out which if any of the many sensory hairs responded to humidity. He found some which did; these hairs produced a nerve impulse faster when the humidity was raised. Kellogg expected that repellents would have some effect on these hairs. So he raised the humidity, added a repellent, and examined the sensory hairs' nerve impulses at the electrodes. The usual response to the increased humidity no longer occurred. It seemed that the repellent had disabled these humidity-sensing hairs.

A similar effect should be observed with heat-sensitive sensory hairs. But no-one knows about this, because the mosquito's heat-sensitive hairs have not yet been found.

Further research continues, in British Columbia and elsewhere. Chemists and biochemists are testing thousands of different compounds for repellency, and biologists are looking in more detail into just how repellents work. It turns out that spherical or oval molecules with a molecular mass of around 200 u are most likely to be good repellents, and biologists are looking into the proposition that these molecules are about the right diameter to block tiny holes in the sensory hairs, so that carbon dioxide or water molecules cannot enter.

Everyone hopes that the results of all this research will make people less attractive to mosquitoes in the future. The prospect of such a result certainly makes it likely that this research will continue.

QUESTIONS

1. What are the two functions of the mosquito's proboscis?
2. Why are mosquitoes sometimes dangerous?
3. What do mosquitoes do when the level of carbon dioxide in the air is first raised?
4. If the level of carbon dioxide is left for a while at this new higher level, what do the mosquitoes do?
5. What characteristics of a hiker walking though a forest on a hot summer day would a mosquito find attractive?
6. Describe what the mosquito does when the hiker approaches.
7. Which parts of the mosquito are used to detect the approach of a target suitable for biting?
8. How does mosquito repellent work?

ANALYSIS

Each of the statements in questions 1 – 6 gives an example of one of the *Ten Principles*. For each, state which of the principles is involved, and explain.

1. Research on topics such as mosquito repellents attracts money from governments and private industry. As a result, rather a large number of scientists work in these areas.
2. Scientists have always known that *something* attracts mosquitoes to people, whether it be smell or something else.
3. The biologists at the British Columbia Research Council hoped and believed that they could find out just what attracts mosquitoes.
4. Biologists had previously thought that people gave off a smell which mosquitoes recognized. When efforts to find the smell failed, they realized that the idea might have to be dropped.
5. "There was no need to postulate a special skin odour as an attractant," commented biologist R.H. Wright. Wright was pleased that his explanation used only the factors already understood: temperature and humidity.

6. Biologists don't know the reason why some chemical compounds repel mosquitoes and other similar ones don't. They are confident, though, that there is a reason.

Questions 7–12 concern the work done by Philip Daykin and his colleagues. This work is found under the title ''Carbon dioxide.''

7. What was Daykin's *hypothesis*?
8. What *experiment* did he and his colleagues do to test this hypothesis?
9. What *prediction* followed from the hypothesis about the result of the experiment?
10. What was the result of the experiment?
11. What additional *hypothesis* is described in the last paragraph of the ''Carbon dioxide'' section?
12. What additional *experiment* did Daykin and his colleagues do to test this new hypothesis?
13. The results of this experiment were in accordance with a *prediction* that followed from the hypothesis. What was this *prediction*?

14. Analyze the work of R.H. Wright and his team in the same way that you used for questions 7 – 13. This work is under the titles ''Warmth and moisture'' and ''How do mosquitoes find you?'' Use the following headings: data base, hypothesis or hypotheses, prediction(s), experiment(s) performed, experimental results, and how these results affect confidence in the hypothesis or hypotheses.
15. Analyze the work of T.E. Kellogg in the same way that you used for question 14. This work is under the title ''How repellents work.''

12 How Cholera Spreads: The Work of John Snow (I)

Cholera is one of the most dangerous diseases a person can get. It produces severe diarrhea, vomiting, and muscular cramps. Dehydration rapidly follows, and the sufferer usually dies within hours of the first attack.

Cholera is virtually unknown in western countries now, but it was much dreaded in the last century. The disease was much more common in poor areas, especially in urban slums. Noticing this, the wealthy and privileged drew comforting moral lessons from cholera outbreaks. Their confidence in their own superiority was reinforced. They were immune from the disease because they were morally superior to the poorer people. ''The disease is rioting in the haunts of infamy and pollution,'' said one newspaper report. ''Drunkards and filthy wicked people of all descriptions are swept away in heaps,'' said another. A just but severe God was often seen as the source of the disease: ''. . .an infinitely wise and just God,'' commented one American state governor, ''has seen fit to employ pestilence as one means of scourging the human race for their sins.'' ''The cholera,'' said a Sunday school newspaper, ''is a scourge, a rod in the hands of God.''

Others, including probably the people who contracted the disease, thought that it struck at random, for no particular reason. But the English physician and medical researcher John Snow, working around 1850, was not satisfied with explanations of this kind. ''It is quite impossible,'' he wrote, ''that even a tenth part of these cases of consecutive illness could have followed each other by mere coincidence, without being connected as cause and effect.'' If this connection could be found, people could be protected. Snow set himself the task of finding out just how the disease was transmitted from one person to another. He first set about learning as much as he could about the history of the disease.

Cholera had first been observed in India, and had come to England in the ships which travelled between the two countries. ''In extending to a fresh island or continent,'' wrote Snow, ''it always appears first at a seaport.'' It was clear, then, that the disease was carried by human beings. Snow made an investigation of individual circumstances of the people who got it. In one of a great number of cases that Snow investigated, for example, a man

> *. . . came home with a bowel complaint, of which he died in a day or two. His death took place on August 18th. His mother, who attended on him, was taken ill on the next day, and died on the day following (August 20th). There were no other deaths from cholera registered in any of the metropolitan districts, down to 26th August, within two or three miles of the above place.*

There were various views about how the disease spread. One was the idea of ''effluvia,'' vapours or odours supposedly given off in the breath of the ill person. Snow's investigations caused him to reject this idea. He noted that not everyone who was in contact with the patient became infected. In particular, if cholera spread this way, doctors would be at great risk. Yet doctors seemed strangely immune. Also, the disease sometimes broke out in new areas, where no sufferer had been. Of one such case, Snow wrote:

> *Whilst the surgeons were vainly endeavouring to discover whence the disease could possibly have arisen, the mystery was all at once, and most unexpectedly, unravelled by the arrival in the village of the son of the deceased John Barnes. . . . He informed the surgeons that his uncle's wife (his father's sister) had died of cholera a fortnight before that time, and that, as she had no children, her wearing apparel had been sent to Monkton by a common carrier. The clothes had not been washed; Barnes had opened the box in the evening; on the next day he had fallen sick of the disease.*

Snow's idea was that the victim's clothes and bed linen

> *nearly always becomes wetted by the cholera evacuations, and as these are devoid of the usual colour and odour, the hands of persons waiting on the patient become soiled without their knowing it; and unless these persons are scrupulously cleanly in their habits, and wash their hands before taking food, they must accidentally swallow some of the excretion, and leave some on the food they handle or prepare.*

Using this idea, Snow was able to explain how cholera went from one place to another, even though the sick person didn't go anywhere. He was also able to explain why the disease didn't hit the doctors who examined the patients—''this being a duty that is necessarily followed by the careful washing of the hands; and it is not the habit of medical men to be taking food on such an occasion.'' And finally, it was easy to see why poor people were more likely to get cholera:

> *It is amongst the poor, where a whole family live, sleep, cook, eat and wash in a single room, that cholera has been found to spread when once introduced, and still more in those places, termed common lodging-houses, in which several families were crowded into a single room.*

QUESTIONS

1. What are some of the symptoms of cholera?
2. Poorer people tended to get cholera much more often than richer people. In the opinion of the richer people, why was this?
3. Why did Snow think that cholera was carried from one country to another by human beings on ships?
4. Why did Snow think that cholera was not carried by "effluvia" given off in the breath of someone who had the disease?
5. How did Snow think that cholera was transmitted?
6. How did Snow explain the fact that poor people were much more likely to get cholera than rich people?

ANALYSIS

1. One early belief about cholera was that it is one of God's ways of punishing sinful people. Which of the *Ten Principles* does this belief go against? Explain.
2. Others thought that cholera strikes at random and for no particular reason. Which principle does this belief go against? Explain.
3. Snow disagreed with both these views. What was his *hypothesis* about how cholera spread?
4. Outline two facts from Snow's *data base* which did not fit in with the "effluvia" *hypothesis*—that cholera spread through the breath of someone who had the disease.
5. Summarize the facts from Snow's *data base* which he used in support of his own *hypothesis*.
6. Snow's own *hypothesis* was able to explain a large number of facts about the disease. Which principle is this feature of his *hypothesis* an example of? Justify your answer by stating three facts which the *hypothesis* explained, and show how the *hypothesis* explained these three facts.

13 How Cholera Spreads: The Work of John Snow (II)

Cholera is a bacterial disease, and the bacteria can be seen under a powerful microscope. They are too small to be seen, though, under the microscopes of Snow's time. What was called the "germ" theory of disease came only with the work of Louis Pasteur and Robert Koch, two decades after Snow's death in 1858. But, as described in the previous article, Snow had gathered convincing evidence that cholera was transmitted from one person to another by means of human excrement. There was, however, one problem with this explanation.

If the cholera had no other means of communication . . . it would be continually liable to die out accidentally in a place, for want of fresh victims.

The theory did not seem to explain the great epidemics, which could strike down enormous numbers of people in one part of a town, while leaving a nearby part of town almost untouched. Snow, though, had gathered a great amount of miscellaneous information about the disease, and it was from these seemingly unrelated occurrences that Snow's next major idea arose. In one case, sewage had been seen flowing into a well, and almost everyone who used the water got cholera. People in an adjacent building used another well, and escaped the disease almost completely. This was in London, and the same thing had been seen in another outbreak, in Manchester. Snow learned, too, of another instance in Ilford, Essex:

The refuse which overflowed from the privies and a pigsty could be seen running into the well over the surface of the ground, and the water was very fetid; yet it was used by the people in all the houses except that which had escaped cholera. That house was inhabited by a woman who took linen to wash, and she, finding that the water gave the linen an offensive smell, paid a person to fetch water for her from the pump in the town, and this water she used for culinary purposes (cooking) as well as for washing.

Incidents of this kind led Snow to an extension of his original idea about how cholera spread:

I allude to the mixture of the cholera evacuations with the water used for drinking and culinary purposes, either by permeating the ground and getting into wells, or by running along channels and sewers into the rivers from which entire towns are sometimes supplied with water.

Snow was well equipped to ask the right questions in future cholera outbreaks. If he was right, it should turn out that people who drank water from one well or pump would get the disease, while people who got their water from somewhere else would escape.

One way for Snow to test this idea would be for him to pollute the water in a well himself and then observe what happened to the people who took their water from it. This, of course, was out of the question. Snow had to wait until a new outbreak occurred, and record who got the disease and who did not. Then he had to find out just where the victims and the non-victims obtained their water. Finally, he could advise people about which well or pump to avoid. If his advice was taken, the disease should die down.

With his investigation planned in advance, therefore, Snow was in a good position to make a systematic study of the next major cholera outbreak. Events in London were to give him his opportunity. In the summer of 1849, there broke out

the most terrible outbreak of cholera which ever occurred in this kingdom. . . . Within 250 yards of the spot where Cambridge Street joins Broad Street, there were upwards of 500 fatal attacks of cholera in 10 days. The mortality in this limited area probably equals any that was ever caused in this country, even by the plague; and it was much more sudden, as the greater number of cases terminated in a few hours.

Snow acted at once.

As soon as I became acquainted with the situation and extent of this irruption of cholera, I suspected some contamination of the water of the much-frequented street-pump in Broad Street, near the end of Cambridge Street; but on examining the water, on the evening of the 3rd September, I found so little impurity in it of an organic nature, that I hesitated to come to a conclusion. Further inquiry, however, showed me that there was no other circumstance or agent common to the circumscribed locality in which this sudden increase of cholera occurred, and not extending beyond it, except the water of the above mentioned pump. . . . On proceeding to the spot, I found that nearly all the deaths had taken place within a short distance of the pump. There were only ten deaths in houses situated decidedly nearer to another street pump. In five of these cases the families of the deceased persons informed me that they always sent to the pump in Broad Street, as they prefered the water to that of the pump which was nearer. In three other cases, the deceased were children who went to school near the pump in Broad Street.

If the cholera was being spread by sewage-contaminated water from the Broad Street pump, then people living in the district who obtained their water from somewhere else should avoid the disease. And there were two large groups of people who indeed seemed immune. In the area was a workhouse, where unemployed people lived in dormitories and worked for their rather meagre keep. Almost none of these people got cholera. The second group was made up of the workers at the Broad Street brewery.

> *. . . on perceiving that no brewer's men were registered as having died of cholera, I called on Mr. Huggins, the proprietor. He informed me that there were about 70 workmen employed in the brewery, and that none of them had suffered from cholera. . . . The men are allowed a certain quantity of malt liquor, and Mr. Huggins believes that they do not drink water at all; and he is quite certain that the workmen never obtained water from the pump in the street. There is a deep well in the brewery. . . .*

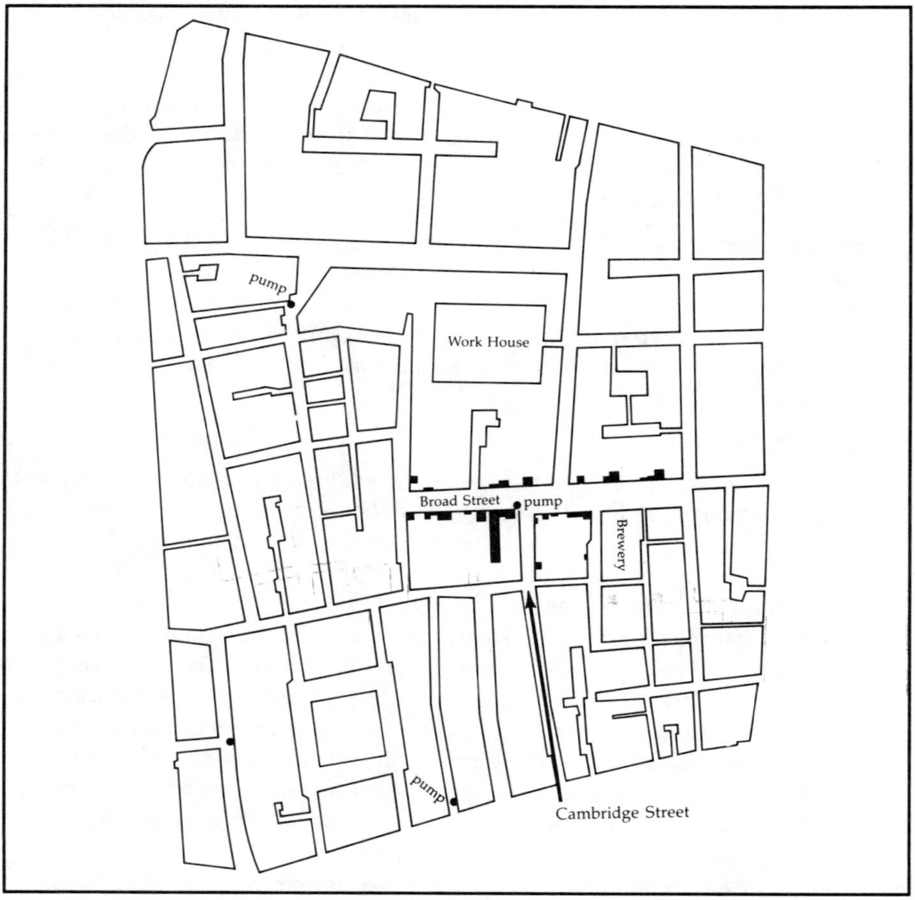

FIGURE 13-1 Map showing area of Cholera Epidemic in London

Snow had the handle on the Broad Street pump removed. He was too late to save the many who had already died, and too late to find out whether removing the handle would arrest the spread of the disease. So great was the suffering that everyone who could move out of the area did so. By the time the handle was removed, the outbreak had already almost died away. Snow again examined the pump water closely through the microscope. There was no direct evidence of leakage of sewage into the water. Snow did observe, though, ''oral animalicules,'' which in his experience were common in contaminated water. And people in the area told him two things they'd noticed about the water just before the outbreak. There was an unpleasant taste, and scum tended to form on the surface when the water was left to stand for a few days.

We now know that cholera bacteria are too small to be seen under the microscopes of Snow's time. But even without direct evidence that the water was to blame, Snow felt that what he had learned was enough to justify his conclusion:

> *Whilst the presumed contamination of the water of the Broad Street pump with the evacuations of cholera patients affords an exact explanation of the fearful outbreak, . . . there is no other circumstance which offers any explanation at all, whatever hypothesis of the nature and cause of the malady be adopted.*

QUESTIONS

1. Cholera bacteria can be seen through a microscope. Why did Snow not observe them?
2. Snow had already gathered a good deal of evidence that cholera was transmitted by human excrement. What facts did this idea fail to explain?
3. What was Snow's explanation of these facts which you wrote about in question 2?
4. Describe the incidents which Snow heard about, which led him to the explanation you described in question 3.
5. When Snow learned of the 1849 outbreak of cholera in London, what did he do?
6. What facts did Snow find which led him to the conclusion that the Broad Street pump was responsible for spreading the cholera?
7. What did Snow do in order to stop more people from getting cholera?
8. This action was not particularly useful, either in preventing more people getting the disease, or in showing whether the pump was in fact spreading it. Why not?

ANALYSIS

1. Snow's original *hypothesis* had been that cholera spread through human excrement. What features of cholera epidemics did this *hypothesis* fail to explain?
2. How did Snow *modify* his *hypothesis*?
3. Describe some of the *data base* Snow used to arrive at his modified version of the original hypothesis.
4. What *prediction* did Snow make about what would be revealed in future investigations of cholera outbreaks?
5. Snow's *experiment* involved extensive inquiries into what water was used by people who got cholera. Describe some of the evidence Snow found to support his hypothesis.

6. From his hypothesis, Snow had made the *prediction* that well water from an afflicted area should contain organic pollutants. Why did he not observe the cholera bacteria?

7. Snow made a *prediction* about what would happen when the handle of the offending pump was removed. What was this *prediction*, and why was it not fulfilled?

8. Snow looked past the terrible suffering in the Broad Street epidemic and concentrated on his investigations. Explain how this aspect of his work demonstrates the Principle of Objectivity.

9. Give an example from the article of the Principle of Tentativeness.

10. Give an example from the article of the Principle of Social Limitation.

14 How Cholera Spreads: The Work of John Snow (III)

After Snow's findings about the dangers of polluted water supplies became known, people who lived in cities began to be more careful about the water they drank.

In London in 1852, there were two private companies which supplied water to the city. One of these — the Lambeth Company — moved its waterworks further up the river, to a place free of sewage. The other company, the Southwark and Vauxhall Company, stayed where it was. Both companies delivered water to the same districts of the city. Snow saw an opportunity for a large-scale experiment:

> *The pipes of each company go down all the streets, and into nearly all the courts and alleys. A few houses are supplied by one Company and a few by the other. . . . Each Company supplies both rich and poor, both large houses and small; there is no difference either in the condition or occupation of the persons receiving the water of the different Companies.*
>
> *. . . no experiment could have been devised which would more thoroughly test the effect of water supply on the progress of cholera than this. . . . The experiment, too, was on the grandest scale. No fewer than 300 000 people of both sexes, of every age and occupation, and of every rank and station, from gentlefolks down to the very poor, were divided into two groups without their choice, and in most cases without their knowledge; one group being supplied with water containing the sewage of London . . . the other group having water quite free from such impurity.*

Many people did not know which company supplied their water, which obviously presented a difficulty for Snow's planned experiment. But he found that the two companies' water had a different appearance, and that he could very easily perform a conclusive chemical test to find which was which. One company's water had a good deal of chloride in it. This had nothing to do with transmitting cholera, but Snow could test for chloride by adding a silver nitrate solution to the water. The water containing chloride became cloudy. In this way, he could find which company supplied the water to any tap in London.

The following summer, another cholera outbreak occurred, from which Snow could gather the evidence he needed. As before, he collected information about the people who got the disease, this time concentrating on which company their water came from. And the results were conclusive. Snow tabulated the following results:

	No. of houses	Deaths from cholera	Deaths in each 10 000 houses
Southwark and Vauxhall Co.	40 046	1263	315
Lambeth Co.	26 147	98	37
Rest of London	256 423	1422	59

FIGURE 14-1

The results could hardly have been clearer. People who drank water supplied by the Southwark and Vauxhall Company were over eight times as likely to get cholera as those using the Lambeth Company's water. There could now be no doubt in anyone's mind that the water supply played a great role in determining who got cholera and who did not.

Once having shown rather convincingly how cholera spread, Snow was in a position to explain a number of things about the disease which had puzzled people for some time. One was that people living in low-lying districts tended to get cholera more than others living on higher ground. Previous researchers had thought that the level of the soil had something to do with the disease, but Snow pointed out that the lower-lying districts were more likely to have contaminated water supplies.

Another puzzling feature of the disease had been that in Scotland, people got cholera all year round. In England, on the other hand, it was almost completely a summer disease. Snow put forward

> *. . . an explanation, to a great extent, of these peculiarities in the progress of cholera. The English people, as a general rule, do not drink much unboiled water, except in warm weather. They generally take tea, coffee, malt liquor, or some other artificial beverage at their meals, and do not require to drink between meals, except when the weather is warm. . . . Consequently, . . . the cholera is chiefly in winter to the crowded families of the poor, and to the mining population, who, as was before explained, eat each other's excrement at all times.*

In other words, English people were protected from cholera in winter by boiling all their water before drinking it.

Snow made a list of twelve recommendations for preventing future cholera outbreaks. With an understanding of how the disease spreads, the recommendations are fairly obvious. They concern cleanliness, washing of soiled clothes, and advice to people and to cities about how to keep their water supply unpolluted. Snow concluded:

> *. . . by attending to the above-mentioned precautions, which I consider to be based on a correct knowledge of the cause of cholera, this disease may be rendered extremely rare, if indeed it may not be altogether banished from civilized countries.*

The authorities in London accepted Snow's advice. In consequence, there were no further epidemics of cholera in the city. Outbreaks occurred, but they soon died away rather than spreading to great numbers of people, as had happened before. Soon, other towns and cities learned the lesson too. The medical profession was rather slower to respond. But not so many years later, in the 1870s, Louis Pasteur and Robert Koch showed that bacteria were responsible for many diseases. After that, no one could ignore the need for cleanliness wherever people were ill. It seems obvious to us today. It wasn't obvious at all before Snow's time.

QUESTIONS

1. What circumstance, in London in 1852, provided Snow with an opportunity to test his idea that cholera could be carried in the water supply?
2. Describe the experiment that Snow had in mind.
3. One problem was that many people did not know which company their water came from. How did Snow solve this problem?
4. In the following year, an outbreak of cholera occurred. From the statistics he collected during this outbreak, what conclusion did Snow draw?
5. Snow had noticed that in Scotland, people got cholera all year round. In England, on the other hand, it was a summer disease. How did Snow explain this?
6. What recommendations did Snow make for preventing cholera?

ANALYSIS

1. When he found out about the two water companies which supplied water to London, Snow could have performed an *experiment* at once to see if the water supply could carry cholera. (Because this *experiment* would have risked the lives of great numbers of people, Snow did not seriously consider actually doing it.) What would this *experiment* have been?
2. Describe the *experiment* that Snow actually did do using the two water supplies.
3. In what way was Snow's *experiment* different from the one you described in question 1, and from most laboratory experiments?
4. The experiment you described in question 1 was unacceptable for humanitarian reasons. Which of the *Ten Principles* is this an example of? Explain.
5. Snow had already collected a great deal of evidence in support of his hypothesis about how cholera spreads. Despite this, when he learned about the way the two companies supplied London with water, he was willing to conduct a large-scale and difficult experiment to test the hypothesis again. Which of the *Ten Principles* does this provide an example of? Explain.

6. A number of facts about cholera had puzzled people for some time, such as why people living on high ground were less likely to get it, and why it was a summer disease in England, but occurred all year round in Scotland. Snow's hypothesis was able to account for facts such as these. Which of the *Ten Principles* does this aspect of Snow's work provide an example of? Explain.

7. Snow gave advice about how to avoid cholera during an epidemic. If you had lived during that time and could not move away from where the cholera was, what steps would you have taken to avoid getting the disease?

8. Snow was willing to accept that whether a person got cholera or not depended on a number of things, not just one. (You described some of these things in question 7.) Which of the *Ten Principles* does this aspect of Snow's work provide an example of? Explain.

15 Driving and Cannabis

People have known for a long time that drinking and driving don't mix, but studies have shown that in North America, on the average, about 6 per cent of drivers at any one time have consumed alcohol.

This may not seem like a particularly high proportion. Perhaps more disturbing is another statistic. When researchers examine accidents in which someone is killed, it turns out that alcohol is a factor in more than half of them. However, there is a more recent concern: the various other drugs that people take — sometimes just before driving. Studies have indicated, for example, that between 60 and 80 percent of marijuana users admit to driving while under its influence.

A major study

It is in this context, then, that in 1978 – 79, the Traffic Injury Research Foundation of Canada (TIRF) was given government funding to conduct a major study in this area. Researchers examined victims of almost 500 traffic deaths, looking for traces of over 90 chemicals, from aspirin, through tranquilizers like Valium, to illegal drugs such as marijuana, cocaine, heroin, and LSD. The purpose of the investigation was simply to find what substances were present in the victims' bodies. But a quotation from an American scientific paper, approvingly quoted in the introduction to the report of the study, makes it clear that this investigation was more than a shot in the dark.

If one knows that a drug has the potential to impair behaviour and performance skills closely related to the driving task, and that drug is in common use by the driving population, one should look to see if the drug is present in accident-involved drivers.

If this was so, then remains of mood-altering drugs would be found in a large number of the dead bodies from car accidents. The study's results, however, showed otherwise — at least as far as the so-called "hard" drugs are concerned. The researchers found no cocaine or LSD in the bodies of any of the people killed. In fact, of the 90 substances under investigation, only four were present often enough to be considered significant.

Topping the list, predictably, was alcohol, present in 55 per cent of the victims. Next was cannabis, present in such drugs as marijuana and hashish, and detected in 12 per cent of the victims. Third was aspirin, and fourth was diazepam (Valium).

A dubious conclusion

"One in every eight victims had been using cannabis," trumpeted an insurance company advertisement, describing the study's results. "Grass can kill on the highway just as lethally as alcohol can." The unstated but clear implication was that cannabis causes one accident in eight.

Such an interpretation, though, would be an excellent example of jumping to conclusions. Certainly something made all the accidents happen. But nothing about the TIRF study showed anything about what that something was. After all, almost as many of the victims had been taking aspirin. But no-one would suppose that aspirin actually caused the deaths. The study showed only what substances were present. It did not show that any of these substances caused the accidents.

Other research strongly suggests that cannabis is a more likely culprit than aspirin. Studies have shown, for example, that cannabis affects reaction time and such driving-related skills as the perception of space and time. As a 1979 National Highway Traffic Safety Administration report to the United States Congress put it: "Experimental studies have shown that marijuana . . . can have significant adverse effects on skills and other measures of capabilities normally associated with driving performance." The same report, though, shows one reason for not jumping to conclusions. "Some researchers," it says, "have reported that marijuana appears to decrease the level of risk that a driver is willing to take."

Clearly, more information is needed before a clear answer to the problem can be given. Like much research of its kind, the Canadian study posed more questions than it answered.

Where to go from here?

The TIRF study, then, is only a first step in answering the question: does cannabis cause fatal road accidents? Perhaps the next question that needs an answer is this: what proportion of drivers at any one time have been consuming cannabis? Suppose, for example, that the figure is 3 per cent. If this 3 per cent were involved in 12 per cent of the fatal acidents, then a driver using cannabis would be four times as likely to be involved in a fatal accident than a non-user.

A study of the proportion of drivers who have been using cannabis doesn't promise to be easy. Unlike alcohol, cannabis cannot be detected by a simple breath test. And, as Dr. Herb Simpson, Executive Director of TIRF comments, ''It's one thing to stop drivers and ask them to blow into a tube. It's quite another to get blood from them, or ask them to produce a urine sample.'' In our society, it is simply not possible to stop motorists at random and force them to give blood or urine samples.

There are ways of getting around problems of this nature — mostly involving paying the people whose blood or urine is to be used. In a large enough study, there are statistical ways of taking those who decline to participate into account. However, a proposal for another study which would have shown what proportion of drivers were using cannabis was turned down because of the cost.

Yet even if results were to become available which showed, for example, that 3 per cent of drivers use cannabis, and that cannabis is involved in 12 per cent of fatal accidents, it still wouldn't prove that cannabis actually causes any accidents. Perhaps the kind of people who use cannabis would be more likely to be involved in an accident in any case. After all, drivers between 17 and 25 have many more accidents than do people in other age groups. And the 17–25 age group could well include a greater proportion of cannabis users. In addition, it's not often that a traffic accident has just one cause. Usually there are several: other drivers, road and weather conditions, mechanical defects. A combination of these factors causes most accidents, not just one of them.

Facts about cannabis

All this doesn't prove, of course, that cannabis is not a factor in road deaths. Still, no-one should assume from it that cannabis is harmless. It has been shown, for example, to have definite effects on a user's memory. Both male and female users show changes in hormone balances—changes that are most pronounced in younger users. Female users show a higher incidence of menstrual problems. Cannabis smoke contains more cancer-causing agents than tobacco smoke, and users also inhale more deeply.

Problems for researchers

Perhaps more disturbing, though, is the extent to which the long-term effects of cannabis are unknown. Dr. Reese Jones, of the Langley Porter

Psychiatric Institute in San Francisco, puts it this way: ''Not long ago I wouldn't have guessed just how complicated marijuana is. Now we're finding out, and it could only be a matter of time before the dangers become manifest.'' Apart from its psychoactive ingredient, tetrahydrocannabinol (THC), cannabis contains over four hundred chemicals, many of which are not found elsewhere. And when consumed, THC is transformed into about twenty-five other compounds. It will take a long time to see whether any of these is harmful. Comments Dr. Jones: ''After all, it took sixty years of steady cigarette smoking in the United States before doctors realized that there was a connection between lung cancer and cigarettes.''

The research which would be most conclusive in this area would be to administer measured quantities of THC to human subjects under controlled laboratory conditions and then to see what resulted. This research is not likely ever to be attempted. It would involve deliberate exposure of human beings to a substance which there is good reason to suppose would be harmful to them. Such a procedure would be unacceptable to the professional ethics of the researchers, and to society as a whole.

As a result, all the controlled studies have been done on other animals, like rats and monkeys. One team of researchers found that cannabis does indeed cause obvious changes in the brain cells of monkeys. Another showed that pregnant monkeys treated with cannabis were four times more likely to abort or to give birth to stillborn offspring. Research in this area is relatively new, though. We can expect to learn far more as time goes on. Meanwhile, individuals still have choices to make, and governments still have to have some policy on these issues.

The questions, then, are not so easy to answer as they might at first seem. Everything that is learned seems to depend on some other piece of information — information that might not be available. At worst, there might be no acceptable way of obtaining it at all. Even when the results are in, great care has to be taken in drawing any conclusions from them. And the situation changes even as it is studied. To return particularly to traffic deaths, the figures could be changed if cannabis became more readily or less readily available, by new driving laws, or by a host of other factors.

People working in the field, though, are confident that more can be learned. It is also likely that funding for this kind of study will continue. After all, if a way can be found to reduce traffic accidents, a lot of people's lives will be saved.

QUESTIONS

The statements in questions 1–4 are all incorrect in one way or another, or are not justified by the information given in the article. Criticize each of the statements.

1. Drugs like LSD and cocaine cannot cause traffic accidents.
2. Drugs like marijuana have not been proven to be harmful, and are therefore safe to consume.
3. There is no evidence to show that any mood-altering drug can cause accidents.
4. Cannabis has been proven to have caused 12 per cent of all traffic deaths.
5. Imagine that a truck-driver friend tells you that he is going to take up cannabis smoking in a big way. Write a letter to try to convince him to change his mind. Use as many pieces of information as you can, both from this chapter and from your own knowledge.
6. You have been appointed Minister of Highways, and you want to carry out a study to settle the problem once and for all: can cannabis lead to traffic deaths? You have three years to complete the study. Give an outline of the investigation you would propose, and list the extra powers you would need, that Ministers of Highways currently do not have.

ANALYSIS

Questions 1–6 concern the 1978–79 study by the Traffic Injury Research Foundation.

1. The purpose of the investigation was to find what substances were present in the bodies of accident victims. Underlying the impetus for the study, though, was a *hypothesis* concerning the effects of mood-altering drugs. What was the *hypothesis*?
2. What *prediction(s)* might follow from the hypothesis?
3. State some facts that might be part of the *data base* out of which the hypothesis would have arisen. The facts can be taken from the article, or from your own knowledge.
4. What evidence contrary to the *hypothesis* did the study reveal?

5. What evidence did the study reveal which supported the *hypothesis*?

6. Explain why the study's results might be different if it was conducted in 1988 – 89. Which of the *Ten Principles* does your answer provide an example of?

7. Find a quotation from the narrative which provides an example of the Principle of Objectivity.

8. Find a quotation from the narrative which provides an example of the Principle of Causality.

9. Find a quotation from the narrative which provides an example of the Principle of Dynamism.

10. When you have answered question 5, explain why it is that a scientist would be reluctant to state that cannabis has been *proven* to contribute to traffic deaths. Use the Principle of Objectivity to help you.

11. When you have answered question 6, use the Principle of Social Limitation to explain why you would need dictatorial powers to undertake the study you propose.

12. Write a report for the local newspaper of an imaginary accident which had as many *causes* as possible. Show how your account provides an example of the Principle of Relativeness.

16 Airplane Crashes and Birds

The big airplane had almost completed its flight. The passengers had finished their meals and the flight attendants had collected the trays. A final announcement had asked the passengers to made sure that their seatbelts were fastened and their chair backs were in the upright position. It was a routine landing in every way.

Suddenly, it wasn't routine any longer. The airliner ran into a flock of starlings. Birds were sucked into the engines and power was lost. The airplane plummeted to the ground. Sixty-two people were killed.

It happened in October 1960, at Boston's Logan International Airport. The plane was a Lockheed Electra.

Birds are a major hazard to airplanes, particularly on takeoff and landing. They are not scared away by the thunderous din of a takeoff. Indeed, airports often seem very attractive places for birds to feed and rest. Runways can be especially attractive in wet weather. Worms come up out of the grass and crawl on the runways — a veritable feast for birds. Many airports around the world suffer from this problem, and crews have to be sent out to sweep up the worms before they attract great flocks of birds.

Often the runways themselves seem to attract some bird species, like gulls. One group of biologists had the idea that it was the warmth of the black tarmac that was attracting the gulls. Near the airfield, they set up electrically-heated pads. The pads, they hoped, would attract the birds away from the runways. The result was a complete failure. The gulls stayed on the runways. They completely ignored the specially-provided pads.

Another idea was to display dead birds in uncomfortable-looking positions near where flocks of the same species usually appear. Dutch biologists tried this technique at Amsterdam's busy Schiphol Airport. Again, the main problem was with gulls, and the decoys were gulls too. The hope was that the stuffed decoys would be seen by the live birds as a danger sign. The birds would then choose somewhere else to go.

When the birds were set up around the airport, the technique worked. The gulls saw them and flew away. Unfortunately, the stuffed birds soon became soaked with rain and had to be replaced. Worse, the gulls seemed to get used to them, and in a few weeks just as many were coming to the airport. Even as the experiment was being carried out, its effect on the birds was changing.

Moving the stuffed gulls from one place to another at the airport helped, but again, only for a while. It was too expensive to keep moving and replacing the stuffed birds, and not particularly effective.

A similar attempt, at New Zealand's Wellington Airport, was more successful. Some people might have thought that the Dutch experience would make the experiment useless, but the New Zealand biologists thought differently. They were dealing with a different kind of gull, the black-headed gull. The problem was a severe one, because large flocks of these gulls roosted at the airport all at once. This had been their habit in this particular oceanside location even before the airport was built.

The experiment worked. The birds went away, and as long as the stuffed gulls were in good condition, they stayed away. Perhaps, thought the biologists, models of dead gulls might work too. They had some plastic models made and installed at the airport. It turned out that the plastic models kept the gulls away just as well as the dead birds did. Two traditional black-headed gull roosts were eliminated for good.

The situation, though, was now rather puzzling. The same procedure that didn't work in Amsterdam did work in Wellington. The biologists weren't happy with this contradiction. There must be a reason for it, they thought. One possibility was that suitable roosting sites were available for gulls quite close to Wellington Airport. And indeed, when the same thing was tried at Whangarei Airport, in the far north of New Zealand, the results were far less impressive. At Whangarei, the airport is very close to the ocean, and the best resting sites other than the airport are flooded at high tide.

An enormous number of experiments like this have been done, and all kinds of ingenious ways have been tried to discourage birds from coming to airports. Firing shellcrackers (shells with firecrackers in them), playing recordings of bird distress calls, flying trained hawks or falcons over them, or even model airplanes constructed to look like predator birds: all these techniques have been tried. None of them works with all bird species, and birds have an annoying habit of getting used to whatever is being used to frighten them off. And these procedures can be very expensive.

Other important ways of discouraging birds include filling in ditches where they like to bathe and drink, closing down nearby garbage dumps where they feed, and making sure that the grass on the airfield is of the correct length. Even the best grass length, though, seems to depend on the species of bird and the airport's location. At Montreal's Dorval Airport, for example, birds avoid grass longer than 15 cm. But at Toronto International Airport, the birds prefer the longer grass.

There seems to be no easy answers. Biologists and airport managers would like nothing better than to find a simple way of scaring away all birds at all times. Unfortunately, nothing of the kind exists. If a particular species of bird is a problem at a particular airport, then its habits must be studied along with everything that affects the birds: climate, nearby features attractive to birds, food availabilty, predators, and a host of other factors.

Yet this work is being done. Each year, the amount of damage done by birds to aircraft is being reduced. Although there are no easy answers, people working in the field are confident that they can learn more about the birds, and about how to reduce to a minimum the problems they pose to the humans who now share the air with them.

QUESTIONS

1. Why do birds find airports attractive?
2. What method was tried to scare gulls away from Schipol Airport in the Netherlands?
3. State two reasons why this method was not particularly effective after awhile.
4. Why was the bird problem particularly severe at Wellington Airport, New Zealand?
5. What differences were there between the method used at Wellington to deal with the bird problem, and the method used at Schipol?
6. Name four other methods of scaring birds away from airports.
7. Name three general problems with the four methods you described in question 6.
8. Name three methods of discouraging birds from staying around airports which do not involve scaring them away.

ANALYSIS

Questions 1 – 5 refer to the idea of setting up electrically-heated pads to attract birds away from airport runways.

1. What *experiment* did the biologists do?
2. What was their *prediction* about the results of the experiment?
3. Upon what *hypothesis* was the prediction based?
4. From what *data base* did the hypothesis arise?
5. As a result of the experimental findings, how did the biologists' view of the *hypothesis* change?
6. At Schipol Airport, near Amsterdam, biologists displayed dead gulls in uncomfortable positions. What *prediction* did they hope would turn out to be true about the result of this *experiment*?
7. What was the *hypothesis* upon which the Dutch biologists based their prediction?
8. How did the *experiment* actually turn out?

9. At Wellington Airport, in New Zealand, the same type of *experiment* was more successful. What was the difference in the result? Why might the difference have arisen?

10. What *hypothesis* could be set out to explain the results of the experiments at Schipol, Wellington, and Whangarei?

11. Show how the behaviour of the birds at Schipol provides an example of the Principle of Dynamism.

12. Show how the New Zealand biologists followed the Principle of Tentativeness.

13. Explain how the fact that so many people have worked on the problem of birds at airports fits in with the Principle of Tentativeness.

14. Read the final few paragraphs of the article, and show how what is described there fits in with the Principle of Relativeness.

15. Find one other principle which applies to the work described in the chapter, and show how the principle you have chosen relates to the work.

PART THREE

CASES IN PHYSICS

Johannes Kepler *(1571–1630) German astronomer, whose studies of the orbits of planets made him one of the founders of modern astronomy.*

17 The Solar System (1): Kepler and Newton

The German astronomer Johann Kepler was a remarkable man. Born in 1571, he contracted smallpox at the age of 3. The disease left his hands crippled and his eyes weak. These are certainly considerable handicaps for an astronomer, but Kepler had another strength that he shared with generations of scientists after him: brilliance in mathematics.

In Kepler's day, astronomy and astrology were one and the same, both involving detailed observation of the heavens. Kepler was a highly-regarded teller of fortunes, casting horoscopes for generals and emperors. He studied the works of the ancient Greeks like Plato and Aristotle, and tried to make sense of some of their mystic notions. He tried, for example, to find some scientific basis for the predictions made from astrological rules. In this he failed, as have all investigators before and since. Kepler's interest in astrology, though, was only partly based on genuine belief and scientific curiosity. Then as now, a great deal of money could be made by astute exploitation of human foibles. Scientific work paid much less.

In 1598, Archduke Ferdinand of Austria issued a decree expelling all Protestants from the University of Graz, where Kepler worked. Kepler was a Protestant, and despite his many influential friends and admirers in Graz, he had to leave. Perhaps it is as well that he did, because afterwards life became very difficult in Catholic countries for those who, like Kepler, viewed the sun rather than the earth as the centre of the planetary system. A previous chapter in this book about the Italian astronomer Galileo explains why.

At the age of 26, Kepler went to Prague, where he became the assistant to the Danish astronomer, Tycho Brahe. Usually known by his first name only, Tycho was the last and greatest of the naked-eye astronomers; within a few years of his death, Galileo's telescopes would revolutionize the study of the skies. Tycho, though, was a magnificently accurate observer, and he accumulated huge masses of data on the stars and the planets. He was, however, unable to give any satisfactory explanation for his findings.

Tycho died in 1601 and Kepler took over his job as the emperor's mathematician. He also inherited the great volumes of information Tycho had collected. Again looking back to Greek mysticism, he tried to fit the positions of the planets into Plato's five solids: tetrahedron, cube, octahedron, dodecahedron, and icosahedron. Since these were the only regular solids, Plato had argued, the various heavenly bodies would have to fit into their corners. Also, since the heavens were perfect, heavenly bodies must move in perfect curves: circles.

The Greek philosophers did most of their work like this. They relied heavily on thought and their conceptions of beauty, but made few formal observations and did still fewer experiments. This preference made them much better at mathematics than at what we would consider science today.

But Kepler had all the records Tycho had made of his observations. These observations simply would not fit with the notion of the regular solids or of a circular path for the planets. Here Kepler's mathematical genius stood him in good stead.

Concentrating first on Mars, he looked closely at Tycho's measurements of its position in the sky over a number of years. Since a circular path didn't fit, Kepler tried a number of other curves whose properties he knew about. An ellipse, he found, would fit. And the sun was exactly at one focus of the ellipse. The same thing was true of all the other planets then known: Mercury, Venus, Jupiter, and Saturn. Later, Kepler showed that the satellites of Jupiter could also be described in the same way. Planets travel in elliptical paths with the sun at one focus: this was Kepler's First Law. He went further. In his Second and Third Laws, he described the speed of the planets at various points in their orbits, and how long each took to go round the sun. Although he had no good hypothesis to explain *why* these laws were followed, Kepler was able to calculate the exact times at which the planets Mercury and Venus ought to cross the face of the sun, as viewed from the earth. No one had observed anything like this in Kepler's time, but he felt able to make the prediction. The planets, he assumed, would continue to move in their paths in accordance with his Laws.

Path of a planet is in the shape of an ellipse.

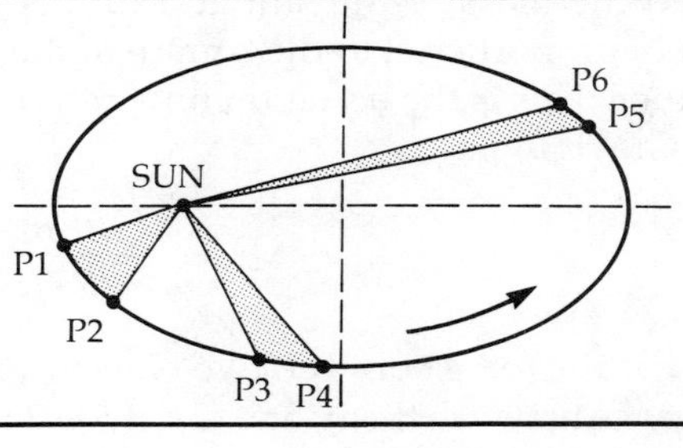

FIGURE 17-1

Less than a year after Kepler's death, the Frenchman Pierre Gassendi watched the sun on one of the days that Kepler had said one of the planets would pass between the sun and earth. Sure enough, within an hour or so of the time that Kepler had predicted, a small dark dot did indeed cross the face of the sun.

Kepler's work described very precisely *what* the planets did. His great achievement was to organize Tycho's work in a different way: to reduce the great mass of Tycho's observations into three simple and elegant Laws. But *why* did the planets move according to these Laws? Some force must surely be guiding them in their paths. But what force? And how? The answer to these problems had to wait until the end of the seventeenth century. It had to wait for an Englishman whom many still regard as the greatest scientist of all time — Isaac Newton.

For almost two thousand years, since Aristotle's time, people had believed that objects in space and on earth obeyed two quite different sets of natural laws. It was Newton's idea that this was not so: the same laws, he thought, should apply to both. In particular, he was convinced that his own First, Second, and Third Laws of Motion should apply in the heavens. A stone being swung round on the end of a string would, according to his First Law, fly off into space — except that the string keeps it going in a circle. Perhaps, thought Newton, a similar argument applies in the case of the moon's motion around the earth. Some force was certainly stopping the moon from flying off into space. Legend has it that Newton was sitting under an apple tree thinking about this question when an apple fell to the ground beside him. It was this, according to the legend, that set Newton to wondering whether the force that pulled the apple to the ground was the same force that kept the moon from hurtling away from the earth. In neither case were strings required.

Whether or not this story is true, Newton did apply his tremendous insight and mathematical genius to the question. And to solve it completely, he had to invent a completely new branch of mathematics. He called it fluxions. We call it calculus.

From the three Laws of Motion, and using his new mathematical methods, Newton was able to prove by rigid mathematical reasoning that Kepler's three Laws would be followed if there was a force between the sun and each planet. The size of this force, he showed, must be proportional to the mass of the sun and to that of the planet. It must also be inversely proportional to the square of the distance between the sun and the planet. This is the equation now referred to as Newton's Law of Universal Gravitation:

$$F = \frac{Gm_1m_2}{R^2}$$

where F is the force between two objects, m_1 and m_2 are their masses, R is the distance between them, and G is a constant called the Universal Gravitational Constant.

The term "universal" is a good one. As well as explaining all three of Kepler's laws, the new formula accounted for the various irregularities in the motions of the planets—something which had really bothered Kepler in his search for perfection in the heavens. These irregularities, Newton showed, were the result of the small gravitational attractions the planets had for each other. Newton was even able to explain the enormously complicated motions of the moon, a problem which had defeated Kepler completely. This was the only topic which, according to Newton, made his head ache.

It was fortunate that Newton lived in England and at the time that he did. Astronomy in Catholic countries was still reeling in the wake of the church's persecution of astronomers Bruno and Galileo; you can read about this elsewhere in this book. To believe that the sun and not the earth was at the centre of things was to invite the attention of the church's enforcement arm: the Inquisition.

Newton had to be persuaded to share his work with the world. To him, it often seemed enough to have solved a problem. Publishing the results often led only to unpleasant criticism and arguments about such things as who had thought of the idea first. And Newton was an irritable and touchy man. He attracted more than his share of controversy.

Fortunately, Newton was a member of an intellectual community which, for the first time since the ancient Greeks, truly encouraged innovative thought on scientific topics like this. There was now a group of people who were eager to hear new ideas and to discuss them in the new coffee houses of the day. For in 1660, when Newton was 34, the Royal Society had been founded in London. The fellows of the Society adopted as their motto the Latin phrase *Nullius in Verba*—words are not enough. To them, it was not enough just to read something written down by one of the ancient Greek philosophers. Careful observation, experimental evidence, and the highest standards of reasoning —these were the only ways to convince this group of people of anything. The Society's members shared a craving for new insights, new information.

It was in this atmosphere, then, that Newton lived. Though he had long been interested in the gravitational question, it took a prod from the outside to get him to publish it. According to a contemporary account, the following conversation took place in 1684, in Newton's house in Cambridge. Present were Newton and Dr. Edmond Halley, after whom Halley's Comet was named, along with the scientist Robert Hooke (of Hooke's Law), and the architect Sir Christopher Wren.

> *After they had been some time together, the doctor asked him what he thought the curve would be that would be described by the planets, supposing the force of attraction towards the sun to be reciprocal to the square of their distance from it. Sir Isaac replied immediately that it would be an ellipsis. The doctor, struck with joy and amazement, asked him how he knew it. "Why" saith he, "I have calculated it." Whereupon Dr. Halley asked him for his calculation without any further delay. Sir Isaac looked among his*

papers but could not find it, but he promised to renew it, and then to send it to him.

This conversation changed the future course of science. Newton did more than he had promised. Three years later, he published his epic *Philosophiae Naturalis Principia Mathematica* (Mathematical Principles of Natural Philosophy). Among many other things, the *Principia* contained Newton's stunningly convincing proof of his gravitational law and the way it explained Kepler's Laws. The book was so comprehensive and elegant a description of the universe, based on so few assumptions and with such powerful mathematics, that it has guided the course of science and mathematics to this day. His work remains a complete and correct account of everything it deals with. It led directly to almost all the great discoveries made in astronomy and physics for the next two centuries and more. Not until 1911, with Einstein's Theory of Relativity, was it so much as suggested that parts of it might not hold up quite precisely or under the most extreme circumstances. And Einstein himself paid this tribute to Newton:

Nature was to him an open book, whose letters he could read without effort. The conceptions which he used to reduce the material of experience to order seemed to flow spontaneously from experience itself. . . . He stands before us strong, certain, and alone: his joy in creation and his minute precisions are evident in every word and every figure.

Newton himself was more modest. ''If I have seen further,'' he said, ''it is by standing on the shoulders of giants.'' There have been many giants since Newton's time. All of them have stood on Newton's shoulders.

QUESTIONS

1. In what topic, besides astronomy, was Johann Kepler interested? List some of his activities related to this topic.
2. Give two reasons why Archduke Ferdinand's decree, which forced Kepler to leave Austria, was probably a good thing for him.
3. What features of Tycho Brahe's work are particularly worthy of respect today?
4. Using Tycho's records, Kepler made a major discovery concerning the five planets then known. What was this discovery?
5. What major question about the orbits of the planets did Kepler *not* answer?
6. What three phenomena was Newton able to explain using the Law of Universal Gravitation?
7. Why is Newton's book, *Mathematical Principles of Natural Philosophy*, still highly regarded even today?
8. What theory finally pointed out some of the limitations of Newton's work?

ANALYSIS

1. Show how the work of Tycho Brahe, Kepler, and Newton fits in with the *data base* and *hypothesis* sections of the *cycle of proof*.
2. The way in which the ancient Greeks viewed the heavens was different from our own. Explain how the Greeks did not follow the Principle of Objectivity or the Principle of Materiality.
3. Without being able to explain why the planets move as they do, Kepler was able to make a *prediction* about two of them. His *prediction* was verified only after Kepler's death. What was this prediction?
4. In making his *prediction* Kepler was applying one of the *Ten Principles*. State what this principle was, and how Kepler used it.
5. Find a quotation from the article which provides an example of the Principle of Causality. Show how the Principle of Causality applies to the situation being described.

6. Find a quotation from the article which provides an example of the Principle of Parsimony. (There are at least three in the article.) Show how the Principle of Parsimony applies to the situation being described.
7. It has been said that Kepler would never have discovered his three Laws but for the actions of the Austrian Archduke Ferdinand. To the archduke they would have represented the gravest error and blasphemy. Explain, using the Principle of Social Limitation.
8. Use the Principle of Social Limitation to explain how Newton was helped in his work by the social framework in which he lived.

Questions 9–14 refer to the following reconstruction of Kepler's sequence of thought which led him to the realization that the planets move in elliptical paths around the sun. (Of course, the work took Kepler years to complete. The actual working out of the mathematics, especially, was long and laborious.)

A. Plato is a man whose work has survived for many hundreds of years.
B. He said that the heavenly bodies move in circles.
C. I think he's right.
D. Here are the paths of the planets over a number of years, which Tycho has left me.
E. I ought to be able to fit these positions into circular paths.
F. I'll try working it out. (Months of mathematics.)
G. No, they won't fit.
H. I'll have to give up the idea of circles.
I. Here is an account of the properties of some other geometrical figures.
J. Maybe the ellipse will fit.
K. If so, these data of Tycho's will fit in with the equation for an ellipse.
L. I'll see if the data do fit in with an elliptical path. (Months of mathematics.)
M. Aha, it works for Mars.
N. The key points of the ellipse are the two foci.
O. Perhaps the sun is at one focus of the elliptical path of Mars.
P. I'll try the mathematics.
Q. It certainly should work out. (More mathematics.)
R. Yes, the data fit. The sun is at the focus.
S. I'll try it for the other planets too.
T. Perhaps they will move in the same kind of path as Mars.
U. It seems to me that they certainly should.
V. As I thought. All the planets move in elliptical paths with the sun at one focus.

Note: After you have completed questions 9 – 13, you should have used each letter A to V once and once only.

9. Find all the *hypotheses* in statements A to V.

10. Find all the *predictions*.

11. Find all the mathematical *experiments* which Kepler performed to test his predictions.

12. Find all the statements referring to Kepler's *data base* which he had before embarking upon this work.

13. Find a statement which shows Kepler *modifying* or *discarding* a *hypothesis*.

14. Find all the statements which show him *successfully verifying* a hypothesis.

15. If we were able to watch Tycho at work, he would have seemed to be performing *experiments* much more than Kepler. Explain why Tycho's years of painstaking and precise observation do not fit into the *experiment* section of the *cycle of proof*.

18 The Solar System (II): The Rise and Fall of Newton's Laws

An Experiment to test the Law of Universal Gravitation

Isaac Newton's Law of Universal Gravitation was a huge milestone in our understanding of the world, and particularly of the solar system. There was, however, a gap in Newton's work. Newton had performed no experiment to demonstrate that there did in fact exist a gravitational force between two objects. He showed that the force could explain the behaviour of the solar system: sun, planets, and moons. But the crucial experimental test remained. Could the force actually be shown to exist here on earth?

In 1798, over a century after Newton had published his *Principia,* another English scientist, Henry Cavendish, saw a way of demonstrating in the laboratory that two objects actually did attract one another. Cavendish attached two heavy gold balls to the end of a rod, and suspended the rod from a fine thread. Very close to each gold ball, he placed a larger fixed lead ball. If the gold balls were attracted by the lead balls, the rod on which they were fixed would turn. To detect the turning motion, Cavendish hung a small mirror on the thread. A beam of light falling on the mirror would then be deflected as the mirror turned. This assumed, of course, that the mirror would turn. But Cavendish was confident that it would.

The experiment worked. The force of gravity did indeed attract the gold balls towards the lead ones. The thread turned, and the light moved round on a scale which measured how far it had gone. As Cavendish had suspected, the gravitational force was very small, but he was able to measure

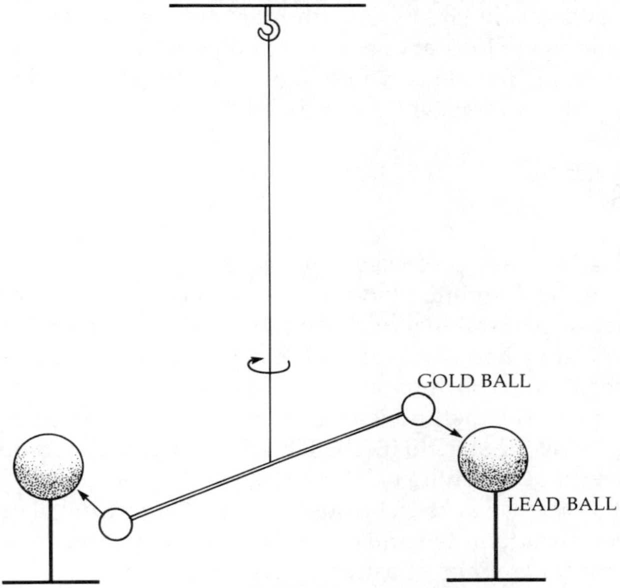

Attraction between the gold and lead objects causes the wire to twist.

FIGURE 18-1

it by finding how big an ordinary force was needed to move the beam of light by the same amount.

Since he knew everything else in the gravitational equation, he was able to calculate a numerical value for the Universal Gravitational Constant, G. Newton's equation could now be written a little differently:

$$F = 6.673 \times 10^{-11} \frac{m_1 m_2}{R^2}$$

The number 6.673×10^{-11} holds true if the force F is in newtons (N), the masses m_1 and m_2 are in kilograms (kg), and the distance R is in metres (m). Cavendish, of course, used the old imperial units in his equation, so the number itself would be different. The number is small, which explains why gravitational forces are so hard to detect — except with very large masses, such as the sun and the planets have. Cavendish was also able to calculate the mass of the earth using his formula. It came to about 6×10^{24} kg, which gives the earth a density of about 5500 kg/m^3 — about the same density as tin and an eminently believable result. If the earth's density had worked out to less than that of hydrogen, for example, or to move than that of lead, it would have been much more difficult to accept Cavendish's findings.

Many people think of scientists as strange, absent-minded, and remote people. Of course, this picture is no more valid than any other stereotype.

But Henry Cavendish must certainly come near the top of any list of odd scientists. Excessively shy even with men, he was so afraid of women that he would never look at one. He left notes around the house to communicate with his female servants. Any woman who accidentally happened upon him was instantly dismissed.

Uranus

William Herschel was a professional musician and an amateur astronomer. He and his sister Caroline made their own telescopes at their home in Bath, England, painstakingly grinding mirrors and lenses for hours on end. By 1774, they had constructed the best telescope in the world, and Herschel began to publish paper after paper describing his observations with it. His favourite method was to choose a section of the sky and systematically list all the stars in it. The Herschels eventually completed four detailed surveys of the entire northern sky, carrying out careful star counts in a total of 3400 small areas. The observations— more detailed than anything previously attempted and of much superior accuracy—were gratefully used by generations of astronomers afterwards.

One night in March 1791, Herschel observed through his telescope an object which appeared as a glowing disk rather than a twinkling point of light like a star. His first thought was that it was a comet, and this is how he reported it. But other astronomers in England and in Europe wanted to find out more. They turned their telescopes on the new object and found it did not have fuzzy edges like a comet. In addition, when enough observations had been completed to compute an orbit, the object was found to move almost in a circle, like a planet. (Comets have very elongated elliptical orbits.) Finally, the orbit was far outside that of Saturn. Now it was clear that what Herschel had seen was the first new planet to be discovered since ancient times. It was the first not to be clearly visible to the naked eye. The planet is now known as Uranus.

Neptune

By 1821, a problem had arisen. In examining the records of Uranus's orbit, the French astronomer Alexis Bouvard noticed that it was not moving around the sun in exactly the way predicted from Newton's Law of Universal Gravitation. It had edged away from where it should be, by about 0.02°. Perhaps, suggested Bouvard, there was another planet, further away than Uranus, that was affecting Uranus's orbit. Such a planet would have to be so far from the sun that it would be extremely dim. It seemed hopeless just to look around for it in the sky.

Another line of attack was to calculate the orbit of the new planet. For data, the way in which Uranus wobbled in its orbit could be used. The

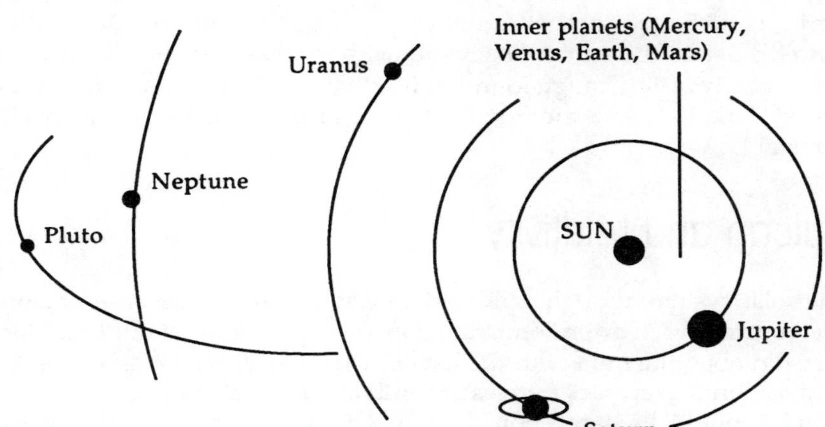

The outer planets. The sizes of the Sun and the planets are not drawn to scale with the planetary orbits.

FIGURE 18-2 The Solar System

calculation would be extremely complicated. Nevertheless, the 23 year-old English mathematician John Adams decided to try it. It took him three years, and then he could not interest astronomers in his solution—perhaps because they did not believe that the calculation was possible.

But it was. Unaware of Adams's solution to the problem, the French mathematician Urbain Leverrier tried it as well, and came up with virtually the same answer. Leverrier wrote to his friend Johann Galle at the Berlin Observatory in Germany. By coincidence, Galle had just made a new and improved star map of the area. On the very first night of searching, Galle found the planet which subsequently became known as Neptune. To scientists and laymen alike, it seemed almost miraculous that the existence of a new planet could be predicted by calculation. Not only did it show that Adams and Leverrier were correct in their calculations, it also removed any lingering doubt about the validity and immense power of Newton's Law of Universal Gravitation, which was the foundation upon which the calculations were built.

Pluto

The Neptune story was repeated, this time by American astronomers. Even with the influence of Neptune taken into account, the orbit of Uranus still wandered off its predicted course. The discrepancy was tiny, but it nagged at Percival Lowell, a wealthy Harvard graduate who had set up his own observatory in the clear desert air near Flagstaff, Arizona. Lowell thought the solution might lie in the existence of yet another planet. He calculated its course in the sky, and his observatory began a long search for it, using better and better telescopes. In 1930, fourteen years after Lowell's death, the search finally met with success. Lowell Observatory assistant Clyde Tombaugh found it. It was named Pluto, after the Greek god of darkness. And the first two letters of its name are the initials of Percival Lowell.

Vulcan and relativity

The solar system, though, still held a mystery. Urbain Leverrier, before even beginning his work on Uranus in 1846, had pointed out that Mercury did not quite fit in with the Law of Universal Gravitation either. Its elliptical orbit precesses (moves around) about 0.012° per century more than it should. His suggestion about how to resolve the conflict between what was observed and what Newton's Laws said ought to be observed will by now come as no surprise. There must, he declared in 1845, be another planet between Mercury and the sun. He called the planet Vulcan. Because it had to be so close to the sun, Vulcan would only be visible during eclipses, and astronomers duly searched for the planet at every subsequent eclipse. There were many reports of its discovery, but all turned out to be mistaken. In the end, the puzzled astronomical community had to accept that Vulcan did not exist. Disturbingly, Mercury's odd behaviour could not be explained at all using Newton's Laws.

The puzzle remained unsolved until Albert Einstein's General Theory of Relativity, published in 1915, supplied the answer. When Mercury was on the other side of the sun, said Einstein, its light rays would be slightly attracted towards the sun by gravity. So Mercury would appear slightly away from its actual position. This idea — that gravity can attract light as well as matter — was by no means the only revolutionary proposition of the theory. Such was the power and depth of thought behind it and the mathematical rigour of its execution, that the theory attracted great interest. Scientists were eager to put the theory to experimental test.

So it was that in 1919, the English physicist Arthur Eddington organized an expedition to Principe, an island off West Africa, where a total eclipse of the sun was due to occur. If the sun's gravity really could bend light, reasoned Eddington, then light from a star which appears close to the sun should be bent. These stars would only be visible during an eclipse, of course. They should, Eddington thought, appear to shift slightly from their nightime positions, when the gravitational influence of the sun on their rays would not be a factor.

Sure enough, the stars near the eclipsed sun did shift their positions. Now at last the problem was resolved. Not only did Vulcan not seem to exist: its existence was shown to be unnecessary.

Einstein's revolutionary Theory of Relativity should not be thought to have disproved Newton's work. On the contrary, the three Laws of Motion and the Law of Universal Gravitation follow directly from the theory. With negligible error, these familiar Laws are followed in all situations where speeds are not too great and forces not too strong. Only under the most extreme conditions do any measurable deviations from these Laws occur: in subatomic particle accelerators, in space near massive bodies, and in similar extraordinary situations.

The Theory of Relativity, though, goes far beyond showing that gravity bends light. At speeds close to that of light, time slows down, distances shrink, and masses increase. These are just some of the predictions of the theory, and experiments with subatomic particles have confirmed them all. The theory occupies a central place in physics and astronomy. Einstein's name is now a household word.

QUESTIONS

1. Give an account of Cavendish's experiment under the following headings:
   a) Purpose
   b) Apparatus—draw a diagram, and explain how the apparatus worked.
   c) Results (qualitative)
   d) Results (quantitative).
2. How does a planet look different from a star, as viewed with the naked eye or through a telescope?
3. When William Herschel first observed the planet Uranus, he thought it was a comet. State two reasons why other astronomers later concluded that it was in fact a planet.
4. What made Alexis Bouvard suggest that there was another planet outside the orbit of Uranus, when no such planet had been observed?
5. During the latter part of the nineteenth century, astronomers all over the world searched near the Sun during eclipses for a new planet which was never discovered. Why did they think such a planet might be found?
6. What revolutionary idea of Albert Einstein's was being tested in Arthur Eddington's 1919 expedition to the island of Principe?
7. Under what circumstances do Newton's Laws of Motion fail?
8. This chapter is headed "The Rise and Fall of Newton's Laws." But Newton's Laws haven't really "fallen." Explain why they are still taught in schools and still used by practising scientists and engineers.

ANALYSIS

1. When Henry Cavendish performed his famous experiment, what *hypothesis* was he testing?
2. What *prediction* did he have in mind about the result of his experiment?
3. In terms of the *cycle of proof*, what was the result of his experiment?
4. If Newton had known that Cavendish would be able to find the size of the Universal Gravitational Constant, do you think he would have been surprised? Explain, using one of the *Ten Principles*.
5. Neptune and Pluto were discovered as a result of observations of Uranus

and the irregularities in its orbit. Analyze the discovery of Neptune in terms of the *cycle of proof*. Identify the *data base*, the *hypothesis* or hypotheses, the *prediction*, the *experiment*, and the result of the experiment in terms of the hypothesis or hypotheses.

6. In the same way as you did for question 5, analyze the discovery of Pluto in terms of the *cycle of proof*.

Questions 7 – 10 refer to the discovery of Neptune and Pluto.

7. Explain how the Principle of Consistency is exemplified by this work.
8. Explain how the Principle of Causality is exemplified by this work.
9. Explain how the Principle of Parsimony is exemplified by this work.
10. Explain how the Principle of Relativeness is exemplified by this work.
11. Even in the 1980s, Uranus's orbit has not been completely explained. Recent computer calculations show that Pluto is too small to account for Uranus's wobble completely. Astronomer Thomas Van Flandern has been investigating possible explanations. "Our best bet," he says, "is for a tenth planet out there." With the aid of a powerful mainframe computer, Van Flandern and his colleagues have made a list of likely places to look.

    Relate this work to as many of the *Ten Principles* as you can.

12. Read the last section of the article, *Vulcan and relativity*, and analyze it in terms of the *cycle of proof*. Two basic *hypotheses* are described; you should deal with each of them in the manner outlined in question 5 above.
13. Explain how Einstein's work and Eddington's subsequent experimental test of it exemplify the Principle of Tentativeness.
14. One of your classmates objects to students having to learn about Newton's Laws on these grounds: "They're out of date, and Einstein showed they were wrong. So there's no point in us learning them." Do you agree? Explain your reasons.
15. Because of the space programs of the U.S.A. and the U.S.S.R., a great deal more is known about the moon and the planets now than was the case even as recently as 1975. Enormous amounts of money have been spent on these projects.

    In terms of the Principle of Social Limitation, explain why this exploration is justified.

    or:

    In terms of the Principle of Social Limitation, explain why this exploration should be terminated.

19 Red Giants, White Dwarfs, Neutron Stars, and Black Holes

Death of the sun

What will happen to the sun when it runs out of fuel? That is a question that has intrigued people for a long time. According to the most recent estimates, though, there's no immediate cause for concern. The sun should continue to shine, about as brightly as it's doing now, for the next five billion years or so.

But other stars have run out of energy, and physicists have made mathematical calculations to see what happens when they do.

According to the calculations, the sun has been shining in much the same way as now for about five billion years. It produces its energy by nuclear fusion — the combining together of two hydrogen nuclei to make a helium nucleus, as in a hydrogen bomb. Eventually, not much hydrogen will be left, and the immense gravitational forces in such a large object as the sun will cause its core to collapse. According again to the calculations, this collapse should result in an increased flow of energy to the surface, which will cause the sun to expand to perhaps fifty times its present size. Being larger, it will pour out more energy, scorching the planets and boiling away our oceans. Stars in this stage of their existence are called *red giants*.

The red giant stage, however, will not last very long. Soon the sun will cool, throwing out a huge amount of gas called a *planetary nebula*. In the centre of this nebula will be the remainder of the sun, which will have shrunk to a few millionths of its present size. The density, of course, will be very high. If a centimetre cube could be dug out from the dead sun, its mass would be about a tonne.

No-one can be sure that all of this really will happen to the sun. But there are stars which appear like this in the sky. These small, very dense masses are called *white dwarfs*. They are surrounded by a glowing planetary nebula.

Neutron stars and exploding stars

Even more interesting is what happens to larger stars as they run out of energy. According to the astronomers' calculations, they explode and throw off a large nebula. What remains in the middle of the nebula is even denser than a white dwarf. Instead of our centimetre cube having a mass of about a tonne, it would have a mass of about a hundred million tonnes. The gravitational forces would be very strong. A human being would weigh about 2500 million times what he does on earth. If one of these stars fell on the earth, it would absorb the whole planet within a few seconds. The earth, shrunken to a small fraction of its present size, would become part of the star.

Because of what astronomers think it's made of, this type of star is called a *neutron star*. The existence of neutron stars was first proposed in 1933 by American astronomers Fritz Zwicky and Walter Baade, of the California Institute of Technology. The theory was more fully worked out in 1939, by the American physicist J. Robert Oppenheimer, later to be director of the Los Alamos Laboratories in New Mexico, where the first nuclear bomb was made. According to their calculations, when the thermonuclear furnace that keeps a large star shining runs down, it behaves differently from a star the size of our sun. The gravitational forces are big enough to collapse the star's atoms. The electrons and the protons combine, and the collapsed star is almost entirely composed of neutrons. As the star collapses, much of its mass is thrown off into space, forming a nebula. In addition, there is a great outpouring of radiant energy: radio waves, light, and x-rays. And indeed, a *supernova*, or exploding star, was observed in 1054 by Chinese astronomers. The supernova, according to their account, was extremely bright, visible even in the daytime. It lasted only for a few weeks, but the remaining nebula can still be seen. It is called the Crab nebula because of its shape.

A few other supernovas have been recorded by the astronomers of the past: Tycho's Nova in 1572, Kepler's Nova in 1604, and others which the historians of astronomy are less sure about. These were all stars in our own galaxy. Modern telescopes are able to look at other galaxies too, and astronomers have recorded hundreds of supernovas. Often, the supernova is as bright as all the rest of the stars in the galaxy put together.

Black holes

According to astronomers' theories, if a star is approximately three times larger than our sun—and many are—it will not form either a white dwarf or a neutron star when its energy runs out. The gravitational forces that make it collapse will be so great that the core of the star will go on collapsing for ever. It then becomes a *black hole*. Once inside a black hole, nothing can escape, not even light. The black hole therefore cannot be seen. And because it traps light, nothing can be observed from behind it: hence its name. Other objects, though, can be in orbit around it.

Robert Oppenheimer had worked all this out in 1939, using Einstein's General Theory of Relativity and some sophisticated mathematical techniques. For the next thirty years, no black hole was observed. There were stars, though, which seemed to be in orbit, but in orbit around an object which could not be seen. Astronomers believed that something must be making these stars travel in circles. But what was it? They did have some ideas. But they didn't really know.

Then it became possible to use the x-ray region of the spectrum to extend the astronomers' view into space. The earth's atmosphere blocks x-rays, so they cannot be studied from the ground. But in 1970, the first x-ray satellite, Uhuru, was launched. Among the first discoveries was an x-ray source, calle Cygnus X-1, very close to one of the mysteriously orbiting stars. Astronomers interpret this observation as being due to a black hole, with an ordinary star in orbit around it. Gas from the ordinary star falls into the black hole, throwing out x-rays as it goes. Observations made on the ordinary star's orbit indicate, too, that the ordinary star's unseen companion has a mass of about four suns—over the three-sun limit previously worked out.

Astronomers are not yet entirely convinced that Cygnus X-1 really is a black hole. One expert on the subject, American Kip Thorne, is quoted as being "about 90 per cent certain that in the centre of Cygnus X-1 there is indeed a black hole."

Nobody is prepared yet, though, to go further than that. Perhaps by the time you read this book, the question will be resolved.

QUESTIONS

1. Write brief point-form notes on the *red giant* under the following headings:
   a) What kind of dying star forms it
   b) What it looks like when viewed from earth
   c) Its density
   d) How it is formed
   e) How long it lasts
2. Write brief point-form notes on the *white dwarf* using the same headings that you used in question 1.
3. Write brief point-form notes on the *neutron star* using the same headings that you used in question 1.
4. Write brief point-form notes on the *black hole* using the same headings that you used in question 1.
5. Describe the relationship between the neutron star, the supernova, and the nebula.

ANALYSIS

1. Oppenheimer's 1939 *hypothesis* was that there are such things as black holes. What *prediction* follows from this hypothesis?
2. Thirty years later, it became possible to perform an *experiment* which produced evidence that the *prediction* was correct. What was this experiment?
3. What was observed in the *experiment* you described in question 2?
4. Find from the chapter a quotation which illustrates the Principle of Causality.
5. Find from the chapter a quotation which illustrates the Principle of Tentativeness.

20 Pulsars

The first pulsar

In the 1930s, American astronomers Baade, Zwicky, and Oppenheimer used Einstein's General Theory of Relativity to show that neutron stars should be formed when supernova explosions occur. But are they? In fact, do neutron stars exist at all? For years, astronomers looked for neutron stars, but failed to find any.

In 1967, a new type of radio telescope was completed by radio astronomer Antony Hewish, at the University of Cambridge, England. The radio telescope was unusual because of its large size (about 18 000 m^2), because of the long wavelength of the radio waves it was designed to detect, and because of the fact that it made measurements continuously instead of averaging them out over a period of time, as most previous radio telescopes had done. On November 28, 1967, Hewish and his research student Jocelyn Bell found an object that was behaving in a most unusual and puzzling way. It was pulsing—emitting regular pulses of radiation every 1.3 s or so. Hewish's first reaction was one of scepticism. "I could not believe," he recalls "that any natural source would radiate in this fashion." The pulses always showed up at the same time of day—not the ordinary day, though, but the astronomical day, which is four minutes shorter because of the earth's revolving around the sun. This showed Hewish that if man-made interference was causing the pulsing signals, they would probably be from other astronomers, who would be the only people likely to do the same thing every 23 h 56 min.

He therefore asked his colleagues at other observatories whether they had any equipment in operation which would produce this type of interference. The answer was no.

FIGURE 20-1 Radio telescope used by Anthony Hewish.

FIGURE 20-2 Radio astronomer Anthony Hewish.

Little green men

The pulses were very different from anything that had ever shown up before. A dramatic new possibility presented itself: perhaps these pulses were the first messages from another civilization, far away from the earth. As Hewish said, ''We had to face the possibility that the signals were, indeed, generated on a planet circling some distant star, and that they were artificial.'' The pulsing source was provisionally named LGM, for ''little green men.''

The idea of little green men, though, is a difficult one for an astronomer to take seriously. Hewish put it this way: ''I would prefer an inanimate explanation . . . in the sense that you can publish it like normal science.'' Hewish knew, too, that the first reports of extraterrestrial life would attract a great deal of media attention, and it would be impossible for him to get on with his work. ''You can't do science with television reporters around the lab,'' he says. ''And you can't keep them out when they come.''

Hewish knew that if the pulses were coming from another planet, the planet would be revolving around a star, just as the earth revolves around the sun. A phenomenon called the Doppler shift would then cause the frequency of the pulses to vary with time as the planet circled its parent star. In the next few weeks, Hewish made measurements which would show whether or not this was happening. He says of this time: ''I felt compelled to maintain a curtain of silence until the result was known with some certainty. Without doubt, these weeks in December 1967 were the most exciting in my life.''

In the event, the results of the Doppler investigation proved negative. The only Doppler shift was due to the motion of the earth around the sun. Meanwhile, research student Jocelyn Bell continued to look for additional pulsing stars. By February 8, 1968, when Hewish and Bell sent their paper to the periodical *Nature*, they were fairly certain that they had found three more. The idea of little green men was abandoned. It would be too great a coincidence for four extraterrestrial civilizations to be sending similar signals.

Why the pulses?

There had to be another explanation, thought Hewish, for the pulses. Two possibilities came to mind: orbiting, or rapid expansion and contraction.

The first possibility, that one object was in orbit round another, seemed impossible. Calculations showed that the fastest possible orbit around a white dwarf would be much too slow; the time for one revolution could not be less than a few minutes. An orbit around a neutron star, on the other hand, would become faster and faster as time went on. The satellite would emit energy rapidly, and would soon collapse back into the neutron star.

The second explanation for the pulses, rapid expansion and contraction, seemed more promising. Other stars do expand and contract, although far more slowly. The fastest pulse every few hours; for most, the period is several days. A white dwarf could behave in this way, but calculations showed that its period would be about a minute — too long. A neutron star doing the same thing, on the other hand, might just pulse fast enough. And this was the idea that Hewish put forward as a possible explanation for his discovery in his February 1968 paper.

Hewish had discovered something completely different from anything previously seen in the sky. His work attracted a great deal of attention from other astronomers. ''I didn't think people would believe it,'' he recalls, ''which of course they didn't. They had to go and check it.'' So all over the the world, radio telescopes were quickly adjusted to be able to pick up signals from these pulsating stars, or *pulsars* as they were now called. By 1974, over 130 pulsars had been logged by radio telescopes in England, Australia, the U.S.A., and the U.S.S.R.

Rotating neutron stars

Meanwhile, the theoretical astronomers also became interested in pulsars. Within a short time, American Thomas Gold put forward a very compelling suggestion about how they work. Rather than expansion and contraction, he suggested, pulsars would be better explained by supposing that the neutron star was rotating, or spinning. The rotation would be very fast — once every 1.3 s, in the case of Hewish's original pulsar. If the star emitted radio waves in some directions and not others, then an observer on Earth would see pulses, in the same way that the rotating lamp in a lighthouse is seen as pulsing by someone a long way from it. Only a very dense object, tightly held together by enormous gravitational forces, could rotate as fast as this. Anything else would disintegrate at once.

Gold's explanation made sense to astronomers. The next step was to see if it stood up to being tested by experiment.

Two important experiments

Up to this point, pulsars had been observed only by radio telescopes. Astronomers with ordinary optical telescopes wanted to see if they could find them too. If so, they would be able to pinpoint the pulsars much more precisely, because radio telescopes are not very precise in finding the exact direction of stars.

Where should they look? If Gold was right, then pulsars are neutron stars. And if Zwicky and Baade had been right in their 1933 calculations, then a neutron star is formed when a supernova occurs.

The radiotelescope had detected one pulsar in the constellation Taurus, where the Crab nebula is. This nebula is the one associated with the 1054 supernova that the Chinese astronomers had recorded. Astronomers at Lick Observatory, California looked through their optical telescope at the Crab nebula. There was one problem, though. The pulsar they were looking for pulsed 30 times per second; it was the fastest pulsar found to date. Such fast pulses would be seen as continuous light by the eye — the principle on which movie and television pictures work. So the astronomers constructed an ingenious piece of equipment. In front of the telescope, they put a set of rotating blades, like fan blades. These blades interrupted the view 30 times per second. If one of the stars in the nebula was indeed the pulsar, then it should be possible to set the blades to make that star disappear. It worked! All the stars but one looked the same whether the blades were in operation or not. One disappeared, revealing itself as a pulsar flashing at 30 times per second, just as the astronomers had expected. This was important backing for the idea that pulsars are neutron stars.

If Gold's rotation idea was correct, another consequence followed. As the neutron star gave off energy, its own rotational energy should go down. It should therefore rotate more slowly as time went on, and the pulses should become slower as well. Gold's calculations showed that the slowing down wouldn't be very much: only a fraction of a percent. But the pulses had been timed over a period of a year or so. And indeed, they did slow down, at exactly the rate that Gold's mathematics had shown they should.

Questions for the future

Work continues. So profound has been the effect of the discovery of pulsars that Hewish was awarded the Nobel Prize for Physics in 1974. The very regular pulses, originating so far from earth, have been used to test Einstein's General Theory of Relativity. (It passed.) There are many ideas about exactly how pulsars give off their radio waves. And the internal structure of a neutron star, so very different from anything here on earth, is of great interest to theoretical physicists.

Some puzzles remain. By 1982, radio astronomers had detected over 300 pulsars, and about 100 supernova remnants had also been seen. In only two certain cases, though, have these two objects been found together. It could be that the pulsing mechanism is so directional that the light from most pulsars misses the earth altogether. On the other hand, pulsars almost certainly last for much longer than supernova remnants. Future work should clarify the relationship between pulsars and supernovae. Or perhaps someone else will come up with a better explanation for pulsars than Gold's. As Hewish says, though, ''There can be no question that pulsars have made, and will continue to make, a considerable impact on the development of astronomy.''

QUESTIONS

1. In what three ways was Hewish's 1967 Cambridge radio telescope unusual?
2. On November 28, 1967, Hewish and Bell found an object in the sky that was behaving in a most unusual way. Name two unusual characteristics of this object.
3. What did Hewish do to see if the signals were coming from a planet that was revolving around a star, rather than from the star itself?
4. One possible explanation for the pulsed signals was that they came from one object in space that was revolving around another. Why did Hewish reject this explanation?
5. What explanation did the American astronomer Thomas Gold put forward for the way in which pulsating stars worked?
6. In looking at distant stars, what advantages do optical telescopes have over radio telescopes like the one Hewish was using?
7. Astronomers at Lick Observatory, California, made an ingenious modification to an optical telescope and used it to look at stars in the Crab nebula. What was this modification and why was it needed?
8. What historical event had made the Crab nebula especially interesting to the American astronomers?
9. In what way was one of the stars in the Crab nebula different from the others?
10. What other property of pulsars turned out to fit in well with Thomas Gold's theoretical predictions?

ANALYSIS

1. In the 1930s, Zwicky and Oppenheimer put forward a *hypothesis* on the basis of Einstein's General Theory of Relativity. What was the *hypothesis*?
2. A *prediction* followed from this hypothesis which was not verified for thirty years. What was the *prediction*?

3. What was Hewish's first *hypothesis* to explain the pulsing signals his radio telescope showed?

4. What *experiment* did he do which resulted in this first hypothesis being discarded?

5. The next *hypothesis* was that of little green men (LGM). Hewish was more than usually hesitant to publish this hypothesis. Which of the *Ten Principles* made him hesitate?

6. The results of two *experiments* made Hewish and Bell discard the LGM hypothesis. What were these experiments and how did they turn out?

7. What was the *hypothesis* which Hewish proposed in the February 1968 paper to explain how pulsars pulse?

8. Astronomers at Lick Observatory, California, performed an *experiment* using their optical telescope. What was this *experiment*?

9. What was the *prediction* that the Lick Observatory astronomers were testing?

10. What *hypothesis*, proposed by Thomas Gold, was the basis for the *prediction* you described in question 9?

11. What other *prediction* followed from Gold's hypothesis?

12. Find a quotation from the article which illustrates the Principle of Tentativeness.

13. Find a quotation from the article which illustrates the Principle of Causality.

14. Find a quotation from the article which illustrates the Principle of Materiality.

15. Find a quotation from the article which illustrates the Principle of Continuous Discovery.

PART FOUR

CASES IN CHEMISTRY

Dmitri Mendeleev *(1834–1907) Russian chemist, whose study of the periodic table of the elements enabled him to predict the existence of elements prior to their actual discovery.*

21 Mendeleev — The Man Who Organized Chemistry

Chemistry must have been a difficult subject for nineteeth-century students. By 1866, over sixty elements had been discovered. Quite a lot was known about most of them and students had to learn about each element separately. No-one had found any good way of organizing all this information.

It was at this time that the Russian chemist Dimitri Ivanovitch Mendeleev returned to St. Petersburg (now Leningrad) from his studies and research in western Europe. Mendeleev saw the need for a good up-to-date chemistry textbook in the Russian language and set about writing one. He was dissatisfied with textbooks that had previously been published in other languages. They were little more than lists of properties of various elements and compounds — a great deal of information, but hard to make any sense of. Mendeleev wanted some way to organize his book. He decided to begin with the elements with the lowest atomic mass, carrying on to end with the heaviest ones. He set up a card file, with one card for each element. Each card had all the properties of the element listed on it.

Mendeleev noticed that if the elements were arranged in the form of a table rather than a list, elements with similar properties would fall underneath one another. Lithium, sodium, and potassium, for example, are all soft metals that react with air and water, so that they have to be stored under oil. They share a number of other properties, too, such as a valence of 1.

These relationships are rather easy to see for the lighter elements, and had in fact been written about before. Earlier attempts to organize the elements into a table had fallen down with the heavier elements, though. Mendeleev was the first to realize that metals like chromium, iron, and copper could not logically be put under any of the lighter elements. Earlier chemists like the Englishman John Newlands had done this in an attempt

Group →	I	II	III	IV	V	VI	VII	VIII
Higher oxides and hydrides	R_2O —	RO —	R_2O_3 —	RO_2 H_4R	R_2O_5 H_3R	RO_3 H_2R	R_2O_7 HR	RO_4 —
1	H(1)							
2	Li(7)	Be(9.4)	B(11)	C(12)	N(14)	O(16)	F(19)	
3	Na(23)	Mg(24)	Al(27.3)	Si(28)	P(31)	S(32)	Cl(35.5)	
4	K(39)	Ca(40)	– (44)	Ti(48)	V(51)	Cr(52)	Mn(55)	Fe(56), Co(59), Ni(59), Cu(63)
5	[Cu(63)]	Zn(65)	– (68)	– (72)	As(75)	Se(78)	Br(80)	
6	Rb(85)	Sr(87)	?Yt(88)	Zr(90)	Nb(94)	Mo(96)	– (100)	Ru(104), Rh(104) Pd(106), Ag(108)
7	[Ag(108)]	Cd(112)	In(113)	Sn(118)	Sb(122)	Te(125)	I(127)	
8	Cs(133)	Ba(137)	?Di(138)	?Ce(140)	—	—	—	
9	—	—	—	—	—	—	—	
10	—	—	?Er(178)	?La(180)	Ta(182)	(W(184)	—	Os(195, Ir(197), Pt,(198), Au(199)
11	[Au(199)]	Hg(200)	Tl(204)	Pb(207)	Bi(208)	—	—	
12	—	—	—	Th(231)	—	U(24)		

**Automatic weights are shown in parentheses. Horizontal lines indicate "missing" elements which Mendeleev expected to be discovered in the future.*

FIGURE 21-1 Periodic classification of the elements (Mendeleev, 1872)*

to keep things simple. Mendeleev realized that it was just too simple. Another category would have to be made for these elements. Nowadays we call them the transition metals.

Mendeleev had two more big problems. First, some elements had properties which placed them in the wrong order according to their atomic mass. Tellurium and iodine are one such pair, as you will see when you do questions 6-10. The atomic masses put iodine before tellurium, the properties put tellurium first, then iodine. Mendeleev resolved this apparent contradiction by saying that properties are what count; an error must have been made by whoever had done the experiments to find the atomic mass. The idea of atomic mass had been established for only a few years, and the 1865 values were still unreliable.

Mendeleev's other big problem was what eventually made him famous. The elements that we now call gallium, scandium, and germanium had not yet been discovered. Yet, in order to make his table work, he had to leave spaces in the positions where we would now place these elements. Mendeleev boldly announced that his table had gaps in it because elements existed which had not been discovered. He went further. He knew the properties of the elements surrounding the gaps in his table. He used these to predict the properties of the undiscovered elements. Here are some excerpts from his predictions about one of these elements, translated into English from the January 7, 1871 issue of the *Journal of the Russian Chemical Society*.

> *Ekaboron should be a metal with an atomic volume of about 15, . . . the specific gravity of the metal should be close to 3.0, . . . its atomic weight 45. The metal will be nonvolatile . . . It should not decompose water at ordinary temperature, but at somewhat raised temperature it should decompose it . . .*

it will dissolve in acids. Its chloride, . . . should be a volatile substance . . . ekaboron chloride wiill be a hygroscopic body, and will be able to evolve hydrogen chloride without having the character of a hydrochloride. . . . its specific gravity will probably be about 2.0. Ekaboron oxide . . . should be a nonvolatile substance and probably should not fuse; it should be insoluble in water, . . . but it will probably dissolve in acid. . . . Anhydrous ekaboron oxide will have a specific gravity close to 3.5. . . .

There was much more. No one could accuse Mendeleev of being vague. His work was greeted by other chemists with great interest, but also with considerable scepticism. In the nineteenth century, Imperial Russia had the reputation of being a primitive and superstitious country, and some scientists, not reading Mendeleev's reasoning, dismissed his predictions as typical Russian mystical rambling.

But in 1875, the French chemist Leroc de Boisbaudran discovered the first of the elements whose existence had been predicted by Mendeleev. De Boisbaudran named the element gallium, after the Latin name for France. Gallium turned out to have exactly the properties that Mendeleev had predicted for one of his missing elements, ''eka-aluminum.'' In 1879, the Swedish chemist Lars Frederick Nilson found another element, which he called scandium, after Scandinavia. Scandium's properties bear a close resemblance to those of Mendeleev's ''ekaboron.'' Finally, ''ekasilicon'' (germanium) was discovered in 1886 by Clemens Alexander Winkler, a German chemist.

Suddenly, Mendeleev was one of the most famous chemists in the world. He continued his scientific and teaching work in St. Petersburg until 1890, when he resigned his academic post in protest about the way students were being oppressed by the Imperial Russian authorities and the university. Although this ended his research, he carefully kept his textbook up to date until he died in 1907.

QUESTIONS

1. Why would chemistry have been a difficult subject for students in the nineteenth century?
2. When Mendeleev began to write his textbook, in what way did he intend that it would be different from previous chemistry textbooks?
3. John Newlands had previously tried to organize the chemical elements into a table according to their properties. For which of the elements was Newlands's table satisfactory?
4. For which of the elements was Newlands's table unsatisfactory and why?
5. In trying to arrange the elements in a table according to their properties, Mendeleev ran across a major problem. His way of solving this problem contributed much to the eventual acceptance of his ideas by chemists. What was the problem and how did he solve it?

Questions 6–10 deal with the charts shown as figure 21-2 and figure 21-3.

	symbol	at. mass	valence	M.P.	B.P.	D
Antimony	Sb	121.8	3 or 5	631	1380	6.7
Arsenic	As	74.9	3 or 5	(sublimes)		5.7
Bromine	Br	79.9	1	−7	59	3.2
Iodine	I	126.9	1	113	184	4.9
Selenium	Se	79.0	2, 4, 6	217	685	4.8
Tellurium	Te	127.6	2, 4, 6	450	990	6.2

(M.P. = melting point in °C; B.P. = boiling point in °C;
D = density in g/cm^3)

FIGURE 21-2 Properties of some of the elements

	As 74.9	Se 79.0	Br 79.0	....
....	Sb 121.8	I 126.9	Te 127.6	

The dots show that there are other elements whose atomic masses lie between 79.9 and 121.8.

FIGURE 21-3 Arrangement of some elements in order of their atomic mass

6. Make a chart similar to figure 21-3 but showing the valences of the six elements, As, Se, Br, Sb, I, Te. Use the information from figure 21-1. Put the elements in the same places, but instead of the atomic mass of the element, write its valence.

7. Make a similar chart to show the melting points (M.P.) of the six elements.

8. Make a similar chart to show the boiling points (B.P.) of the six elements.

9. Make a similar chart to show the densities (D) of the six elements.

10. What change did Mendeleev make in the arrangement of the elements from the way they are set up in figure 21-3? Use your answers to questions 6–9 to explain why.

ANALYSIS

1. What *predictions* did Mendeleev make about discoveries that would be made in the future?

2. Upon what *hypothesis* did Mendeleev base these predictions?

3. What did Mendeleev use as the *data base* for his hypothesis?

4. Later work *strengthened confidence* in this hypothesis. What work was this?

5. What feature of chemical knowledge before Mendeleev's time violated the Principle of Parsimony?

6. Mendeleev's contribution to chemical knowledge did not lie in his discovery of new chemical facts. Rather it lay in his pointing out the *relationships* between facts already known. Which of the *Ten Principles* is being pointed out by the statement above? Explain.

7. Mendeleev abandoned his scientific research at St. Petersburg in 1890. Which of the principles does this incident provide an example of? Explain.

8. In a few of his predictions, Mendeleev was wrong. He expected, for example, that more careful research would show that iodine's atomic mass is greater than that of tellurium. (To see why, look back at your answer to questions 6 - 10.) But by 1913, the atomic masses of the elements had been much more accurately determined. The values are as shown below:

I	Te
126.9	127.6

So Mendeleev was mistaken. The elements arranged in order of *properties* are not quite in order of *atomic mass*. Chemists in 1913 didn't think any less of Mendeleev for making this wrong prediction. They accepted, though, that Mendeleev's conclusion was not in line with the new evidence and had to be changed. Which of the *Ten Principles* does this incident provide an example of? Explain.

9. Chemists in 1913 were still disturbed by the exceptions in the smooth upwards order of atomic masses in the periodic table. They would have liked things to be simpler, with no exceptions. Which of the *Ten Principles* does this preference provide an example of? Explain.

10. In 1913, the English chemist Henry Moseley was using X-rays to investigate the charge on the atomic nucleus. (We now know that this charge is the number of protons in the nucleus; we call it the atomic number.) When the elements are arranged in order of atomic number rather than of atomic mass, the irregularities in the Periodic Table which are discussed in question 19 disappear. The result is the Periodic Table as we see it today. The properties of the elements do fit if the elements are arranged in order of atomic number.

    Mendeleev had thought he had the answer to the problem. He turned out to be wrong. Forty years later, in 1913, the problem was finally solved. Further, Moseley's work led to a lot more research on the atomic nucleus itself.

    This account of Moseley's work points out another of the *Ten Principles*. Identify the principle, and explain.

22 Nuclear Chemistry (I): The Discovery of Nuclear Fission

In the late 1930s, many physicists were working with the atomic nucleus, trying to unlock its secrets — to find out exactly what it was and how it behaved. One of the most powerful tools was the neutron, discovered in 1932 by the English physicist, James Chadwick. The neutron has no charge, so it can be shot at the nucleus of an atom without being repelled. In fact, the nucleus sometimes absorbs the neutron to form a new, heavier nucleus. Many of these artificially-made nuclei are unstable, and themselves give off other particles that are radioactive. Sometimes another element can be made in this way. Ordinary aluminum 27, for example, absorbs neutrons to become aluminum 28. But this new nucleus soon emits an electron. Because electrons are so light, the mass of the nucleus is hardly changed at all. But the charge does change. When the electron, which has a negative charge, is emitted, the positive charge on the nucleus increases by one, from 27 to 28. Of course, the new nucleus is no longer aluminum. It is the next element along in the periodic table — silicon.

This type of work was being done in various laboratories around Europe in the late 1930s. An interesting question occurred to the Italian scientist Enrico Fermi: what happens if neutrons are fired at the heaviest of all the elements: uranium? Uranium has 92 protons; what should happen, reasoned Fermi, is that a new element with 93 protons should be produced. This would be a completely new element, the first to be made by man. This indeed would cause a sensation among the scientists of the time.

Fermi reasoned that the new element should come under rhenium in the periodic table. He knew what kind of properties rhenium had, and therefore what the new element should be like. He bombarded pieces of uranium with neutrons, and found traces of new radioactive substances. Try as he might, however, he could not get the experiment to work as he expected. No trace of anything resembling rhenium could be extracted from the neutron-bombarded uranium.

Another pair of scientists, Otto Hahn and Lise Meitner, working in Berlin, had been trying the same thing, and with a similar lack of success. It was 1938, and since Meitner was Jewish, she had to leave Hitler's Germany. Hahn carried on his experiments with the German physicist Fritz Strassman, and they found that one of the products of the reaction had properties very like those of barium, but was radioactive. They concluded that what was being formed was radium, just below barium in the periodic table. To prove this, they tried to separate the radium from the barium. They could not do it, and neither could another team from France. These scientists were working with some of the best chemists in the world, and they found it strange and frustrating not to be able to succeed. The chemical separation ought to be fairly easy, after all.

Now a refugee in Stockholm, Lise Meitner suggested a solution to the riddle: the new substance wasn't just *like* barium; it *was* barium. The idea would certainly explain why it couldn't be separated from barium. Of course, it had to be a different isotope from those which occur naturally. That would explain its radioactivity.

Another problem: barium has a much smaller atom than uranium. It has only 56 protons, compared with uranium's 92. How, then, could it be formed? Meitner suggested the answer. The uranium atom, she reasoned, must have split, almost down the middle, to form two much smaller atoms. One of these would be the barium that Hahn and Strassman's experiments kept suggesting.

Within a few weeks, scientists from all over the world had dropped everything to test this sensational idea that a heavy nucleus can be split into two new ones of about half the size. Success at last! Barium *was* one of the products. The atom had been split. To this breaking-up of a large atom into two roughly equal pieces, Meitner gave the name fission.

Hahn received the Nobel Prize in 1944 for splitting the uranium atom, and in 1966, Hahn, Meitner, and Strassman were co-winners of the prestigious Enrico Fermi Award.

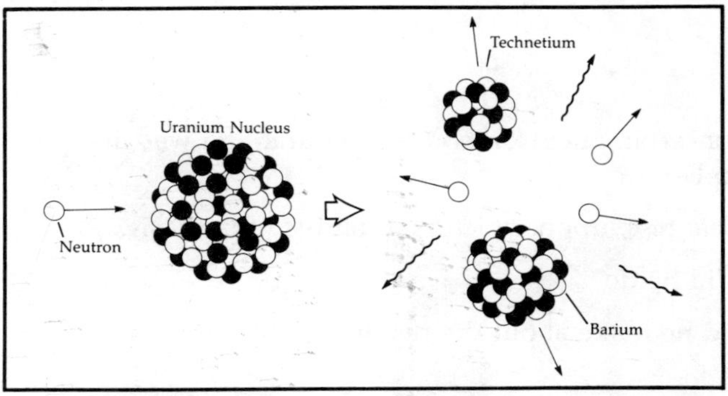

FIGURE 22-1 Uranium Fission

QUESTIONS

1. What property of the neutron allows it to be shot at an atomic nucleus without being repelled?
2. When an atom absorbs a neutron, what happens to its atomic mass?
3. When an atom emits an electron, what happens to its atomic mass?
4. When an atomic nucleus emits an electron, what happens to its atomic number — the positive charge on the nucleus?
5. Enrico Fermi fired neutrons at uranium nuclei. What did he expect to happen?
6. When Fermi actually did fire neutrons at uranium nuclei, what happened?
7. Otto Hahn and Lise Meitner also fired neutrons at uranium nuclei. What two facts did they discover about one of the products of this nuclear reaction?
8. What did Hahn and Fritz Strassman think this product was?
9. What feature of this product did Hahn and Strassman find mystifying?
10. What did Lise Meitner, now working in Stockholm, suggest as a possible solution to this mystery?
11. List the names of all the scientists in the narrative. For each scientist, write one line to summarize the scientist's contribution to the discovery of nuclear fission.
12. Which scientist do you think should be named as the discoverer of nuclear fission? Justify your answer.

ANALYSIS

1. What *hypothesis* concerning neutron-bombarded uranium was first put forward by Enrico Fermi?
2. Outline Fermi's *data base,* upon which he built his hypothesis.
3. What *experiment* did he do?
4. What *prediction* did he make about the result of his experiment?

5. This hypothesis of Fermi's gives an example of one of the *Ten Principles*. Identify the principle, and show how the *prediction* you mentioned in the previous question fits in with it.

6. How did the *experiment* turn out?

7. Your answer to the previous question should provide an example of another of the *Ten Principles*. Identify the principle, and show how the result of the experiment provides an example of it.

8. Fermi, Hahn, and Strassman couldn't get their experiment to work as they had predicted. What *hypothesis* did Lise Meitner put forward to explain their lack of success?

9. A number of other scientists immediately performed experiments to test Meitner's hypothesis. What was the *prediction* that followed from Meitner's hypothesis.

10. Interestingly, besides barium, the other product of uranium fission turned out be be element No. 43, technetium. Technetium does not occur in nature, and had first been produced only two years before, in the cyclotron at the University of California. Now here it was again, adding some interesting new information to the *data base*. Scientists were pleased to learn more about this new element, but not particularly surprised.

    The information in the previous paragraph provides an example of one of the *Ten Principles*. Identify the principle, and explain why this information fits in with it.

11. Draw a condensed version of the cycle of proof flow chart, including data base, hypothesis, experiment, verification, and the NO and YES loops. Consider the work of Fermi, Hahn, and Strassman. Trace the path around the *cycle* that their work took.

12. Draw another condensed version of the *cycle of proof* flow chart, and trace the path taken by Meitner and the scientists who picked up her hypothesis.

23 Nuclear Chemistry (II): An Influential Letter and A Pioneering Experiment

The discovery of atomic fission in the winter of 1938 – 39 was a major event in the history of science. For over a century, chemists had regarded the atom as the basic building-block of all matter. In 1808, the English chemist John Dalton had come to the conclusion that atoms were indestructible, and it had always been one of the first things taught to beginning students of chemistry. Now this most fundamental of all chemical ideas had to be given up. Under the right conditions, the atomic nucleus itself could be split up into two smaller-sized parts.

Chain reactions

The previous chapter described how Enrico Fermi fired neutrons into a piece of uranium, and how Lise Meitner showed that the uranium nucleus split into two smaller nuclei, barium and technetium. By the spring of 1939, it had become clear that this fission, as it was called, was accompanied by two interesting effects. First, a great deal of energy was released. Second, every time a fission occurred, more neutrons were released, which in turn could cause another uranium nucleus to split up. In fact, each fission produces 2.47 neutrons on average, each of which is potentially capable of splitting up another uranium nucleus if it hits one. The new nucleus, in turn, produces more neutrons. This is the idea of the chain reaction.

Chain reactions are nothing new to chemists. An ordinary fire is a chain reaction. A match sets a small piece of wood alight. Heat from this small piece ignites more wood, and the fire grows. Eventually, unless the fire is put out, all the available wood is burned, and energy is released in the form of heat.

The principle behind the nuclear chain reaction is the same. Nuclear reactions, though, are different in two ways. Each stage of the chain reaction takes place very quickly indeed — a fission takes only about 2×10^{-11} s. And the amount of energy produced is many times greater than for any ordinary chemical reaction. The fission of a gram of uranium produces as much energy as burning 300 L of oil. It was clear that nuclear fission was a discovery which had enormous potential — for good, for evil, or for both.

Problems with uranium

Could a controlled nuclear chain reaction actually be made to happen? Various difficult problems stood in the way. First was the difficulty in obtaining a pure uranium sample. Uranium is not a particularly rare element. It is, for example, 400 times as common as gold. But unlike gold, it does not occur in pure form, and in 1939, no satisfactory method had been developed for purifying it. Up to that time, no-one had been very interested in it; neither the metal nor any of its compounds were of much use for anything except academic study.

The problem of purification, though, was not the most difficult one which had to be solved. As it occurs naturally, uranium always consists of the same mixture of two isotopes: about 0.7 per cent U-235, with 92 protons and 143 neutrons, and the rest U-238, with 92 protons and 146 neutrons. The Danish physicist Niels Bohr, though, had been working on his so-called "liquid-drop" theory of the atomic nucleus, and had arrived at the conclusion that isotopes with odd numbers of nucleons (protons + neutrons) would be more likely to fission than those with even numbers of nucleons. If Bohr was right, then the U-235 would be responsible for all the fission, and therefore all the energy. In addition, only the U-235 would produce the neutrons necessary to make other uranium atoms split up; the U-238 would just be in the way. This U-238 constitutes 99.3 per cent of all the uranium nuclei. If a sustained chain reaction was to be achieved, it looked as though it would be necessary to separate the two isotopes in order to make a pure sample of U-235.

The scientists' fears

Meanwhile, events in the world outside the scientific community began to lend this work a sense of extreme urgency. All through the late 1930s, Nazi Germany had been rapidly arming itself and taking over smaller countries like Austria and Czechoslovakia. On September 1, 1939, Germany attacked Poland. Britain and France had a treaty to defend Poland and two days later they declared war on Germany.

The scientists were alarmed. Many of them, like Lise Meitner, had already had to leave Nazi Germany. Most were horrified by the Nazis, and they feared what would happen if the Germans were able to achieve the chain reaction and use it to make a bomb. A few grams of uranium would do as much damage as hundreds of tonnes of TNT. The most powerful opponents of Nazi Germany, Britain and France, were already frantically at work trying to rearm to counter the German threat. They were unlikely to be able to divert enough resources to start a crash program to get ahead and stay ahead in what appeared to be a race to make a nuclear bomb. The most likely place to persuade the politicians to act vigorously to counter the threat seemed to be the United States.

The most famous scientist of the day was Albert Einstein, discoverer of relativity. His name was a household word, and at the outbreak of war in 1939 he had been living in the United States for six years, working at the Institute for Advanced Studies in Princeton, New Jersey. He seemed the best person to write to President Roosevelt to suggest that the United States make every effort to produce such a bomb before the Nazis did. Somewhat reluctantly, because he was a pacifist, Einstein agreed, and wrote the following letter to Roosevelt:

Sir:

Some recent work by E. Fermi and L. Szilard, which has been communicated to me in manuscript, leads me to expect that the element uranium may be turned into a new and important source of energy in the immediate future. . . . I believe, therefore, that it is my duty to bring to your attention the following facts and recommendations.

In the course of the last few months, it has been made probable—through the work of Joliot in France as well as Fermi and Szilard in America—that it may become possible to set up nuclear chain reactions in a large mass of uranium, by which vast amounts of power . . . would be generated. Now it appears almost certain that this could be achieved in the immediate future.

This new phenomenon would also lead to the construction of bombs. A single bomb of this type, carried by boat or exploded in a port, might very well destroy the whole port together with some of the surrounding territory.

I understand that Germany has actually stopped the sale of uranium from the Czechoslovakian mines which she has taken over. That she should have taken such early action might well be understood on the ground that the son of the German Undersecretary of State, von Weizsacker, is attached to the Kaiser Wilhelm Institute of Berlin, where some of the American work on uranium is now being repeated.

Yours very truly,
A. Einstein.

A pioneering experiment

Whatever President Roosevelt's private opinion of Einstein's letter, it was not possible at that time for him to take the kind of action that Einstein was suggesting—a massive project to develop an atomic bomb. America was not yet involved in the war, and a powerful segment of American public opinion wanted to keep it that way. In Canada, which had declared war on Germany shortly after Britain did, things were different. Canadian scientific and industrial resources were limited at that time, and Canada was working flat out on the immediate task of helping Britain survive the opening moves of the war—the danger of a German invasion of Britain was very real. Nevertheless, Canadian scientists, led by physicist George C. Laurence, were keeping up with the developments in the nuclear fission field. To them, the stakes were very large. "We assumed," said Laurence, "that the German scientists were also hard at work, and we were convinced that if Germany produced a weapon first they would win the war."

It had become clear that there was another problem with getting a chain reaction with uranium to work. The neutrons emitted when fission occurred were mostly too fast-moving to initiate another fission in other uranium nuclei. If the chain reaction was to be set going, the neutrons were going to have to be slowed down before hitting another uranium nucleus. Some other substance—a moderator—was going to have to be mixed with the uranium to do this. The best candidate was heavy water. But, as Laurence later recalled, "heavy water was scarce and costly to produce. The 185 kilograms that the French scientists had obtained from a hydroelectric plant in Norway and brought to England was most of the world's supply." And this heavy water was being used for experiments in England.

Laurence continues:

Rough calculations with the inaccurate data then available suggested that it might be possible to obtain a large release of energy using some form of carbon instead of the heavy water with the uranium. Carbon would be less suitable for the purpose, but was cheaper and easier to obtain. I decided to experiment with carbon and uranium oxide.

Laurence borrowed 450 kg of uranium oxide from Eldorado Gold Mines, who produced it as a waste product from the extraction of radium. He put the uranium oxide into coffee bags, surrounding each bag with larger sacks of powdered carbon. The carbon, recalls Laurence, "was in the form of ten tonnes of calcined petroleum coke, a very fine black dust that easily spread over floors, furniture, and ourselves." The bags and the sacks were placed in a wooden bin a little less than 3 m in diameter and lined with paraffin wax to reflect neutrons back into the reactor rather than letting them escape. In the centre was the neutron source, a piece of radium.

The purpose of the experiment was to determine whether a very large release of nuclear energy would be possible in a large bulk of the kinds of uranium and carbon which I had. It would be possible if at least as many neutrons were released by fission as were captured. That implied that if an independent source of neutrons is surrounded by a small quantity . . . of the combination of uranium and carbon, more neutrons would reach the surrounding walls than if the combination of materials were not present.

Laurence was hoping that his sacks of carbon and his coffee bags of uranium oxide would produce more neutrons on the outside than were fed into them by the piece of radium on the inside. There was no danger of the chain reaction becoming self-sustaining like a fire—a phenomenon now described as a reactor ''going critical.'' Still less could the arrangement turn itself into a bomb. Laurence knew that the amounts he was working with were far too small for that. Also, of course, he was working with natural uranium, containing only 0.7 per cent of the readily-fissionable U-235.

Laurence began his experiments in March 1940. Since he was already working on an urgent war project, this work had to be done mainly in his spare time. He took over a room in the National Research Council laboratories in Ottawa, changed the lock, and kept the only key. Then, with a few colleagues who helped during their vacations, he began to build his experimental reactor.

The experiment was not a success. The reactor did not work. More neutrons were captured by impurities in the uranium oxide, the carbon, and the bags than were produced by fission of the uranium nuclei. Laurence summarized the result in the following way:

. . . our measurements had shown that the release of neutrons by fission in our combination of materials was a few per cent less than capture. Therefore it would not be possible to obtain a large release of nuclear energy in that combination of materials even if large quantities were used. There was too much loss of neutrons by capture in impurities in the coke and uranium oxide and in the small quantities of paper and brass that were present. We did not then realize how a little impurity could lead to failure.

A giant project

By this time, events were moving faster in the Unites States, which was now in the war as a result of the Japanese attack on Pearl Harbor. A giant project, noncommitally named the Manhatten Project, had been authorized by President Roosevelt to manufacture an atomic bomb in the shortest possible time, with expense no object. This was a far cry indeed from

George Laurence, working in his spare time in Ottawa. As it turned out, this massive effort was indeed needed to achieve success before the end of the war. Nothing less would have worked.

QUESTIONS

1. What long-established belief had to be given up when the first nuclear fission reaction was achieved in the laboratory?
2. Explain what is meant by a nuclear chain reaction.
3. In what ways is a nuclear chain reaction different from an ordinary chemical chain reaction, such as the burning of a piece of wood?
4. What problems had to be overcome before the first nuclear chain reaction could be achieved?
5. What is the function of the moderator in a nuclear reactor and why is it needed?
6. Name two substances which can be used as moderators in nuclear reactors.
7. What is meant by a nuclear reactor ''going critical''?
8. What was the purpose of George Laurence's 1940 experiments in the National Research Council laboratories in Ottawa? Why did these experiments not work?

ANALYSIS

Questions 1 – 4 refer to the last paragraph of the letter from Albert Einstein to President Roosevelt.

1. State the *hypothesis* which is being put forward.
2. List the facts in the *data base* which led to the hypothesis.
3. State a *prediction* which might have been made from the hypothesis.
4. Assume that the *hypothesis* was correct. Under what circumstances might the *prediction* still not have been fulfilled?
5. In fact, wartime German work on nuclear fission was quite half-hearted. Hitler considered work in this area as ''Jewish science''; the government put a low priority on it, many people working on it had already fled, and most who remained judged it more prudent to put their efforts elsewhere. Which of the *Ten Principles* is in operation here? Explain.
6. The atom had long been regarded as the fundamental building-block for all matter. Now that it had been shown that atoms could be split,

chemists had to rethink their ideas. Which of the *Ten Principles* were they demonstrating when they did this? Explain.

7. To produce a working nuclear reactor, the materials would need to be pure, in order to avoid neutron capture. There would have to be a moderator: heavy water or carbon. The reactor would have to be big enough, and it was likely that the uranium would have to be enriched. Each of these factors depended upon the others. For example, if large amounts of impurities were present, greater enrichment of the uranium would be needed. Which of the *Ten Principles* does this feature of nuclear reactions provide an example of? Explain.

8. Although no-one had yet made a nuclear chain reaction work, the scientists of the day were confident of being able to learn enough to construct a successful reactor. Which of the *Ten Principles* were they demonstrating when they showed this confidence? Explain.

9. Analyze the experiment done in Ottawa by George C. Laurence in terms of the *cycle of proof*. Describe the *experiment*, the *hypothesis* upon which it was based, the *prediction* about the result of the experiment that followed from the hypothesis, and explain what contribution the work made to the *data base*.

10. Apart from incidents already described in previous questions, find examples from the chapter of the Principle of Social Limitation in operation.

24 Nuclear Chemistry (III): The Atomic Bomb

The first working self-sustaining nuclear chain reaction was achieved in the United States, in a squash court under one of the stands at the University of Chicago football stadium. The reactor, then called an "atomic pile," was much larger than anything previously built: about 10 m by 11 m in area, and over 6 m high.

The uranium enrichment process

By late 1942, experiments had shown that Niels Bohr's original prediction had been correct, and that indeed U-235, which constitutes only 0.7 per cent of natural uranium, was the isotope which actually splits in the fission process. The other 99.3 per cent is U-238, which does not participate in the fission reaction. Quite the opposite: it actually absorbs neutrons. Experiments had shown that natural uranium can be enriched— the proportion of U-235 increased—by a method known as gaseous diffusion. In 1942, as part of the American effort to make a nuclear bomb, construction began of plants to enrich enough uranium to make the bomb work.

Because this work was by now being done under the strictest security, even the city built to accommodate the plants and the people who worked in them was a secret new one: Oak Ridge, in Tennessee. People living in the area, of course, knew that great amounts of material were being delivered to the just-born town, that vast construction was going on, but that nothing, for the moment, was being produced and sent out. Rumours were rife: a nudist colony, a home for pregnant female soldiers, a factory to make Roosevelt campaign buttons or windshield wipers for submarines, or a new American Vatican City — Italy, of course, was fighting on the German side.

The factories represented a giant gamble. There was no guarantee that the process to enrich the uranium would work on anything like this scale. Even if it did succeed in producing enriched uranium in the amounts that seemed necessary, no one could be sure that a bomb made with enriched uranium would actually explode. The scale was truly enormous. The plants cost a billion dollars each to build, and during the war years, they used more electricity than the whole of France. Since not enough copper was available for the wiring, the entire installation was wired with silver, borrowed from Fort Knox.

The idea was that uranium would be combined with fluorine to form uranium hexafluoride, a liquid that can be made into a gas fairly easily. Using this gas, the two uranium isotopes were to be separated using a process called gaseous diffusion. The gas was forced through a series of porous barriers. Uranium hexafluoride molecules containing U-235 atoms are slightly lighter than those with U-238 atoms. As a result, according to Graham's Law of Diffusion, they pass through the porous barriers very slightly faster. The uranium in the the gas coming through the first barrier contains only a little more than 0.7 per cent U-235. Because the enrichment is so slight, the gas has to be passed through the barriers very many times — hence the giant size of the plant.

This, at least, was the intention. It seemed sound enough, but until the plants were built, no one could be sure that the process really would work. Under normal circumstances, staking so much on such a new process without going through a series of pilot projects would be unthinkable. But circumstances were far from normal. If the Germans were to be beaten in the race to make a bomb, speed was of the essence.

Success in the squash court

Meanwhile, back in Chicago, the team led by Enrico Fermi was continuing work at the reactor. Like the Canadian physicist George Laurence, Fermi had calculated that a self-sustaining chain reaction was possible, even without enriched uranium. By now, though, it was clear that the uranium fuel and the graphite moderator would have to be made from the purest possible materials. Fermi was fortunate in that he had about six tonnes of very pure uranium to work with. The carbon moderator was in the form of graphite blocks, which caused certain problems, as Fermi's wife Laura recalls:

> *Everything became black. Graphite powder covered the floor and made it black and slippery as a dance floor. Black figures skidded on it, figures in overalls and goggles under a layer of graphite dust. . . . The carpenters and machinists who executed orders with no knowledge of their purpose, and the high-school boys who helped lay bricks must have wondered at the black scene. Had they been aware that the ultimate result would be an atomic bomb, they might have renamed the court Pluto's Workshop or Hell's kitchen.*

Previous exploratory experiments had enabled Fermi to calculate the approximate size that the reactor would have to be in order to be self-sustaining. Even as the reactor was being built, more experiments were being performed on it so that more precise estimates of the best shape could be made and of the exact size that would be needed.

By the morning of December 2nd, 1942, everything was complete. Fermi himself described the crucial experiment to a visitor:

The pile is not performing now because inside it there are rods of cadmium which absorb neutrons. One single rod is sufficient to prevent a chain reaction. So our first step will be to pull out of that pile all control rods, but the one that George Weil will man.

This pen will trace a line indicating the intensity of the radiation. When the pile chain-reacts, the pen will trace a line that will go up and up and that will not tend to level out. . . . The counters will click faster and the pen will move up to this point, and then its trace will level off. Go ahead George.

The scientists worked very slowly. As the last cadmium rod was being gradually pulled out, they checked the instruments to see that all their calculations were working out correctly. Laura Fermi explains why:

According to the theory, an explosion was out of the question. The release of lethal amounts of radiation through an uncontrolled reaction was improbable. Yet the men in the squash court were working with the unknown. They could not claim to know the answers to all the questions that were in their minds. Caution was welcome. Caution was essential. It would have been reckless to dispense with caution.

At 3.20 p.m., as he had done many times that day, Fermi asked Weil to withdraw the last control rod a little more. This time, however, he turned to the anxious group watching from the balcony. ''This will do it,'' he said. ''Now the pile will chain-react.''

It did. The counters began to click at a great rate. The pen went up and did not level off. A chain reaction was taking place, and the reactor was producing heat at the rate of about half a watt. The age of nuclear energy had begun.

A number of reactors working on the same principles as Fermi's, but of more advanced design, were built at Oak Ridge and at another new city: Hanford, Washington. Their main use was not to produce energy, but rather to manufacture plutonium. The unwieldly plant that had been set up at Oak Ridge was necessary to separate the U-235 from U-238, which captures neutrons. In the new reactors, though, this neutron capture was put to use. When U-238 captures a neutron, it forms U-239. But U-239 itself is unstable, quickly breaking down to form an even heavier element, plutonium. The isotope formed is Pu-239. With its odd number of neutrons (145), plutonium-239 should, according to Bohr's liquid drop model of the nucleus, be a candidate for fission — just like U-235, but more so. And experiments in 1941 had shown that this was indeed so. If anything, plutonium was a better potential material for a nuclear bomb than U-235. Just as U-235 needed an enormous plant, though, to separate it from U-238, so plutonium, not occurring naturally, likewise needed great plants to make it: nuclear reactors like Fermi's.

The first three nuclear bombs

The authorities decided to hedge their bets and make both kinds of bomb, the uranium type and the plutonium type. In the event, both were used. By July 1945, enough uranium had been enriched at the massive Oak Ridge plants to make one bomb. Almost enough plutonium had been produced to make three bombs. That was all there was.

In 1939, Einstein had thought that a nuclear bomb would have to be so big that a ship would be needed to move it into enemy territory. Three years later, physicists and engineers were betting that a bomb could be made small enough to fit into an airplane. If plutonium or enriched uranium were used instead of natural uranium, they reasoned, the chain reaction should sustain itself in a much smaller-sized object than Fermi's reactor. The ''critical mass'' needed for a self-sustaining fission reaction could be calculated. Fermi's successful predictions about the behaviour of his reactor had shown that the theory behind such calculations was on the right lines.

Of course, a bomb could not be controlled and started gradually like Fermi's reactor had been. A different mechanism had to be designed. In the plutonium bomb that was to be the first one exploded, two smaller pieces of plutonium are fired at each other by a high-explosive mechanism so that together they exceed the critical mass. A chain reaction begins and within a short time all the nuclear fuel is expended. The uranium bomb works in the same way, but using enriched uranium rather than plutonium.

In a bomb, the chain reaction must be made to proceed as quickly as possible. Theory had indicated that plutonium should absorb enough neutrons to keep the chain reaction going even with fast-moving neutrons. So should U-235, unless too much U-238 was present to capture the neutrons. With plutonium, therefore, or with sufficiently-enriched uranium, no moderator is needed. The neutrons move faster, the nuclei to be split are closer together and the whole process is quicker. The entire bomb has to be enclosed in a thick casing to hold the uranium or plutonium together long enough for most of it to react without the whole assembly turning to vapour.

These were the basic ideas behind the construction of the first atomic bombs. At another secret new city, Los Alamos, New Mexico, another vast establishment was set up to perfect the design of the two types of nuclear bombs and, as the enriched uranium and the plutonium became available, to build them. Of course, it was impossible to test the bomb on a small scale. Either there was enough fuel to reach the critical mass when it all came together, or not. Only one kind of test was possible.

The bomb explodes

It took place at 5.30 a.m. on July 16, 1945, at Alamogordo, New Mexico. The temperature at the centre of the explosion was 10 000 times that of the surface of the sun. Someone on another planet would have been able to see the bomb go off. The tower upon which the bomb had been constructed literally vaporized. A spectator in Arizona, 250 km away, saw "the sun come up and go down again." The power of the explosion was greater than the scientists had expected. Most of the instruments they had set up around the desert to monitor the effects of the blast were smashed. Physicist Philip Morrison was one of the closest eyewitnesses, 16 km from the bomb:

> *The whole desert pre-dawn suddenly burst into light, which we expected. What I did not expect was the heat from the bomb . . . it was the heat of noon. The face got warm suddenly, and you knew that that heat was coming from the bomb, ten miles away. Then the desert lit up, and [there was] the purple glow, and the mushroom cloud, glowing in the dark, and then a long time later, nearly a minute later, the sound.*

Hiroshima and Nagasaki

Watching the bomb, an American enlisted soldier commented, "Buddy, you just saw the end of the war." The war against Germany was already won. The war against Japan, though, looked likely to last another year and take hundreds of thousands of lives. The decision was made to force the Japanese to surrender by using the other two available nuclear bombs on the cities of Hiroshima and Nagasaki. An ultimatum was sent to Tokyo: surrender, or the bomb will be used. It was rejected as "unworthy of public notice." On the morning of August 6, 1945, the uranium bomb was dropped on Hiroshima and 78 000 people died. Three days later, a second bomb containing plutonium was dropped on Nagasaki. This time there was a fire storm and 100 000 people were killed or maimed. Years later, people were still dying from radiation damage. The bombs achieved their aim. On August 14, the Japanese government surrendered.

It was naive to have supposed, as the Americans did, that the secret of how the bomb was made could be kept for long. On September 22, 1949, the Soviet Union exploded its first nuclear bomb. By 1983, bombs had been made with explosive power equivalent to fifty million tonnes of TNT, over 2000 times the power of the Alamogordo bomb. Over 50 000 nuclear bombs, missiles, and warheads are capable of killing everyone on earth many times over. American, British, French, Chinese, and Soviet scientists, technicians, and engineers all devote their lives to enlarging and improving the nuclear arsenals. At the same time, great numbers of people all over the world are very much opposed to the proliferation of nuclear weapons. In their view, humanity would be better off without this technology.

QUESTIONS

1. Write a short account of the gaseous diffusion process under the following headings:
   a) What the process does
   b) How it does it
   c) Why the process is important
2. Write a short account of Fermi's nuclear reactor under the following headings:
   a) Fuel
   b) Moderator — what it does and why it is needed
   c) Purpose
3. In 1945, two types of nuclear bomb had been constructed. Compare and contrast the two types under the following headings:
   a) Fuel used
   b) Difficulties in extracting or manufacturing the fuel

ANALYSIS

1. Read the account of the first self-sustaining nuclear chain reaction, achieved by Enrico Fermi in 1942. Identify some *predictions* made by Fermi on the day the reactor first operated.
2. State some of the facts in the *data base* used by Fermi as he built the reactor.
3. What effect did the successful operation of the reactor have on the various *hypotheses* Fermi based his predictions upon?
4. The hypothesis that a large enough plant would produce enough enriched uranium to make a nuclear bomb before the end of the war led to the construction of the uranium enrichment plant at Oak Ridge, Tennessee. The chemists and engineers who designed and built the plant had a *data base*. List as many facts as you can from this *data base*.
5. The nuclear bomb and the uranium enrichment process are examples of *technological* rather than *scientific* advances. All the basic chemistry and physics had already been done; what was required was a successful plant to make enough enriched uranium and to build the bomb itself — more of an engineering than a scientific task. In this type of situation, the *hypothesis* and the *prediction* sections of the *cycle of proof* tend to come together. The *hypothesis/prediction* for the uranium enrichment process

is stated in question 4. State a similar *hypothesis/prediction* for the construction of the U-235 bomb.

6. The first reaction of most people who watched the explosion of the first nuclear bomb in the New Mexico desert was that no nation could withstand attack by weapons of this size. Analyze the events of this time and the next few weeks in terms of the *cycle of proof*: a *prediction* which was made at the time of the explosion, *experiments* that were done which tested the *hypothesis*, and the reaction of the Japanese government.

The statements in questions 7 – 11 refer to the *Ten Principles*. State the principle which applies to each of the statements and explain.

7. Large numbers of scientists gave up what they were previously doing and immersed themselves in the development of the nuclear bomb.
8. By 1942, Enrico Fermi knew a great deal about the behaviour of small amounts of uranium and graphite. He expected that the large reactor he made in Chicago would behave in a way that could be predicted from that knowledge.
9. During the early part of the war, much was still to be learned about the behaviour of plutonium. But everyone was confident that enough could be learned and, therefore, a nuclear bomb could be built.
10. An Italian, Columbus, discovered America in 1492. A second Italian, Fermi, was the first to build a successful reactor in 1942. Some astrologers, making much of the interplay of numbers, regarded this as significant. The scientists didn't take the idea seriously.
11. After the war, many scientists were so horrified by the effects of the bomb that they gave up work on nuclear energy and concentrated on something else.
12. People's views vary about who is responsible for the threat to humanity posed by the 50 000 nuclear weapons scattered around the world. Some say the scientists who discovered nuclear fission and the chain reaction are to blame. Others say it is the politicians who have decided that more money is available for work on armaments than for most other scientific and engineering endeavours. Others again point out that nuclear weapons have never been used in war since 1945, and assert that they have preserved the balance of power and therefore prevented war. What do you think? Explain and justify your opinion.

PART FIVE

CASES IN BIOLOGY

Charles Darwin *(1809–1882) English naturalist who first posed the theory of evolution in his book,* The Origin of Species.

25 Evolution (1): Charles Darwin

For a long time, people have known that changes in domestic animals such as dogs, horses, cattle, or cats can be produced by selective breeding. To breed speedy racehorses, for example, breeders carefully select two horses who are themselves fast. Not only this: it's important that all the ancestors of the horses are known as well (the pedigree), and that they too were fast in their day. When two pedigree horses are mated, the resulting foal is likely to have many of their best qualities. If it does, it will be bred in turn with other pedigree horses. By selecting and mating the best horses in each generation, breeders can produce animals which can run much faster than any of their ancestors. This breeding process, carried out over many generations, is called artificial selection.

Artificial selection has also been used to develop more than just faster horses. Before cars and trucks became common in the 1920s, horses were the main means of hauling goods over short distances. And on farms, they performed most of the work now done by tractors. Horses bred for these purposes are very different from racehorses. They are heavier, slower, stronger, more resistant to wounds and diseases, and have a much quieter temperament. Artificial selection, in fact, can produce horses with almost any combination of desired characteristics you care to name. All these artificially-bred animals, of course, are much different from their wild ancestors of several hundred years ago.

Artificial selection has also produced the great variety of pedigree dogs and cats we see at dog and cat shows. It has produced cows that can give much more milk than their ancestors from only a few decades ago. Other cattle breeds are very efficient at turning grass and grain into beef. Present-day grain, vegetable, and fruit crops, too, are the products of intense and determined artificial selections.

The Origin of Species

The English naturalist Charles Darwin was interested in selective breeding, and considered the breeders' achievements in his famous book *The Origin of Species*, first published in 1859. If the various domestic animals and plants could be changed by selective breeding he argued, then perhaps wild animals, too, had come to their modern form by a similar process. It would not, of course, be artificial selection, as in domestic animals, but natural selection. Darwin's theory was that only the best of any breed of animal survive: best in terms of adaptation to the environment, being able to find food, escape from predators, and so on. The rest either starve or fall prey to their enemies. Over a number of generations, the animal changes so as to be better matched to its environment. For example, Darwin explained the evolution of the giraffe's long neck in this way:

> *The giraffe, by its lofty stature, much-elongated neck, fore-legs, head and tongue has its whole frame beautifully adapted for browsing on the higher branches of trees. . . . With the nascent developing giraffe, the individuals which were the highest browsers and were able during dearths or famines to reach even an inch above the others will often have been preserved. These will have left offspring inheriting the same bodily peculiarities, whilst the individuals less favoured in the same respects will have been most liable to perish. By this process long continued an ordinary hoofed quadruped might be converted to a giraffe.*

Finches on the Galapagos Islands

In his famous voyage from 1831 to 1836 in the sailing ship *Beagle*, Darwin observed and recorded in most meticulous form the details of the plants and animals which live on the coasts around South America. What struck him most were the birds and animals on the Galapagos Islands, an isolated group of islands in the Pacific. There he noted no fewer than fourteen kinds of finch, of various sizes, all with differently-shaped beaks, and all feeding on different foods. Some ate seeds; these had larger and heavier beaks than mainland finches. Others ate insects, and had long, sharp beaks. One species picked insects from holes in trees; it had a beak very like that of the common European woodpecker. No other kinds of land bird were to be seen on the Galapagos, which are over 1000 km from the nearest mainland.

Why, wondered Darwin, were the Galapagos finches so different from those on the mainland? And why were there so many kinds? Darwin looked for an answer to these questions. Perhaps, he thought, the finches were all descended from a few ancestors, possibly blown to the Galapagos from the mainland by a violent storm. These original mainland finches would eat seeds, as finches on the mainland still do. Because there were few predators, there soon would be so many finches that not all could find enough seeds to survive. Some would die of starvation, but others would change their diet from seeds to insects and other possible food sources. The most successful insect-eating finches would have different beaks than the seed-eating finches. The population of finches would gradually begin to change, to evolve.

Natural selection

How could this evolution take place? Darwin's explanation did not require the action of any unknown or outside ''guiding hand.'' In his explanation, Darwin first made use of the fact that offspring are not identical to their parents. If two finches with ordinary beaks mate, some of their offspring will have slightly shorter beaks, while others will have slightly longer beaks. Sometimes, one of the offspring will have an exceptionally long beak — considerably longer than either of its parents.

Second was the idea of natural selection. According to Darwin, pressures from the environment help some individuals in a population to survive, and others not to. For example, in the Galapagos, after the first seed-eating finches arrived, there would soon be more finches than could be supported by the available seeds. Perhaps a few finches would have beaks a little stronger than the rest. These would be able to crack seeds which the other finches could not — perhaps seeds which nothing else on the Galapagos could eat. These finches would live longer, and would therefore produce more offspring. The finches in the next generation would have, on the average, slightly larger and stronger beaks. Again, some of these would have offspring with larger and stronger beaks than any of their parents. Over many generations, this process would continue to produce more finches with longer and stronger beaks. A new kind of finch would have evolved.

Darwin made careful observations, too, of many of the other unique Galapagos species. In particular, the giant tortoises amazed him. Similar in many ways to their mainland cousins, they have much longer necks, and some are so big that a person can ride on their backs. As well, tortoises from the different islands are slightly different from each other — the islands are too far apart for a tortoise to be able to swim from one to the next.

A long neck would help the tortoise reach leaves which its shorter-necked relatives couldn't reach. In times of drought, when there would be less food, long-necked tortoises would survive, while the shorter-necked ones starved. After several generations, most of the tortoises would have longer necks. Then, in another drought, only the ones with even longer necks could survive and reproduce.

Large size would also be an advantage to a tortoise when water was scarce. A larger tortoise is better able to store and retain water, and so could survive a longer dry spell. In each generation, more of the larger tortoises would live and produce offspring. Because offspring are likely to have some of their parents' characteristics, they would, on average, be large too. Over a period of many generations, therefore, the entire population of tortoises would have developed longer necks, and would be larger in size than their ancestors.

Social Darwinism

Darwin worked on these ideas from 1836, when he returned to England, until 1858, when he first published his account of his Theory of Natural Selection—survival of the fittest. The theory challenged many ideas from the Bible that had been accepted as factual. As a result, Darwinism caused a great deal of controversy. Nevertheless, the notion of natural selection appealed greatly to the Victorian laissez-faire capitalist ideas of Darwin's time. Many Victorians took the view that it was inevitable and right that the sharpest and most ruthless businessman should become rich, while those who couldn't keep up should fall by the wayside. In the second half of the nineteenth century, there were extremes of wealth and poverty. The poor were very poor indeed, and the rich were extremely rich. Taxes were very low by present standards, and there were almost no social welfare and social security schemes like there are today. Darwin's ideas were applied to human affairs and "social Darwinism" was put forward as a way of justifying the extremes of wealth and poverty.

It is not primarily for these reasons, though, that the idea of evolution was almost universally accepted among biologists very soon after Darwin published *The Origin of Species*. The theory fitted in with the scientific evi-

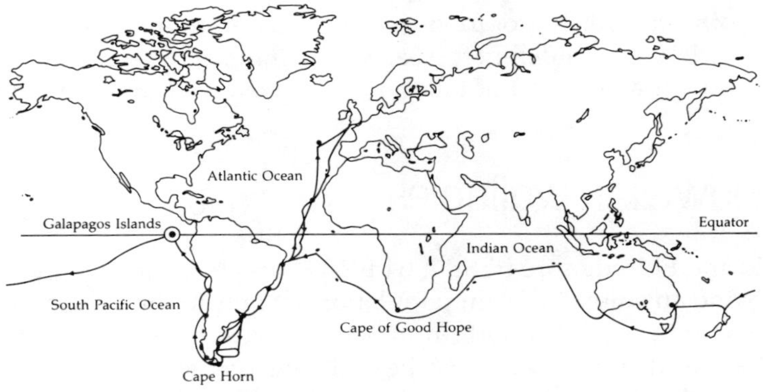

The route of Darwin's trip around the world on H.M.S. Beagle.

FIGURE 25-1

dence that was becoming available during Darwin's time and afterwards. Not only did the theory fit the facts; it also made sense of them and helped organize them. Some of the evidence supporting the theory is outlined in the next chapter.

There was much that Darwin didn't know. Today, for example, we know how genes carry traits from parents to offspring, but none of this had been discovered in Darwin's time. He was well aware of the need to keep an open mind. In his own words:

> *I have steadily endeavoured to keep my mind free so as to give up any hypothesis however much beloved (and I cannot resist forming one on any subject), as soon as the facts are shown to be opposed to it. Indeed, I have had no choice but to act in this manner, for with the exception of the Coral Reefs, I cannot remember a single first-formed hypothesis which had not after a time to be given up or greatly modified.*

Modern developments

Well over a century after Darwin's death, some aspects of the theory of evolution are still in question. In the past few years, for example, some scientists have taken the view that the big steps in evolution — from one species of plant or animal to another — must take place in small groups of organisms. In the view of Stephen Jay Gould and Niles Eldredge, these small groups would be separated from the rest by such things as a newly-formed river or a stand of trees. Or perhaps some natural disaster could kill off most of a population, leaving only a small number behind. Evolution would then take place faster, perhaps in thousands rather than millions of years. In Gould's words: "Major evolutionary change may occur in these small isolated populations. Favourable genetic variations can quickly spread through them." This would explain the fact that a layer — millions of years worth — of fossilized remains is often found just above another thick layer of different fossils, with nothing between the two layers to show how the one kind of plant or animal evolved into the other.

The Hawaiian wallabies

If Gould and Eldredge are right, it would be possible, somewhere, sometime, to find an animal or plant population actually in the process of such rapid evolution. A 1982 report, from field biologist James Lazell, Jr., suggests that exactly this process may be occurring on the Hawaiian island of Oahu. In 1916, a pair of Australian wallabies escaped from a zoo. They survived and reproduced, and by the early 1980s several hundred wallabies lived wild on the island.

The Hawaiian wallabies are considerably smaller and lighter in colour than Australian wallabies. At least one of their liver enzymes has a different amino acid structure. These changes have helped the new species — called the Kahili rock wallaby — to adapt to life on Oahu. The enzyme change, for example, allows them to eat Hawaiian plants which would be poisonous to their Australian cousins.

Lazell came rather reluctantly to the conclusion that rapid evolution had occurred. "At first I suspected we were dealing with a form that had simply disappeared from Australia some time after the ancestral pair had been shipped to Hawaii," he recalls. "But when I went to Australia and checked all the historically documented specimens, nothing matched." Even now, he is not sure that the species really is in fact an example of rapid evolution. It is, after all, possible that the two wallabies which originally escaped were just odd examples of Australian wallabies to begin with.

The scientists disagree

Other scientists doubt this theory of Gould and Eldredge. They argue that just because a species appears suddenly in the fossil record, it does not mean that it evolved suddenly. Fossil sites are few and far between, after all. The absence of fossils could be explained if evolution occurred in some small region where conditions were not suitable for fossil formation. Then the new form might have spread back to the original location, replaced what was there before, and flourished for millions of years more. Then the fossil record which paleontologists discover today would be the same: one layer suddenly ending, another suddenly beginning.

Darwin would not have been surprised by these developments. To use his own words again:

> *A grand and almost untrodden field of inquiry will be opened on the causes and laws of variation, on correlation, on the effects of use and disuse, on the direct action of external conditions, and so forth. . . . In the future I see open fields for far more important researches.*

This particular prediction of Darwin's has certainly proven true.

QUESTIONS

1. What animals do you know of which have been changed by selective breeding?
2. Suppose that people have developed a taste for skunk meat, and gloves made from skunk tails are selling well. A new type of skunk is being produced by selective breeding, to be raised on farms. What skunk features will breeders try to get rid of as much as they can? What features will they try to emphasize?
3. Explain how a new breed of cat with short legs and a long tail could be produced by selective breeding.
4. What is the difference between *artificial selection* and *natural selection*?
5. Using your own words rather than Darwin's, explain how the giraffe's long neck might have developed by natural selection.
6. Explain how the zebra's stripes help the zebra fit into its environment.
7. Explain how the zebra might have evolved from plain, dark-coloured ancestors.
8. In what way did Darwin's theory of natural selection contradict the beliefs of many people in his time?
9. In what way did Darwin's theory appeal to the Victorian laissez-faire capitalist ideas of his time?
10. Often, paleontologists digging for fossils find two layers of fossils of similar but not identical species, one above the other. How would Stephen Jay Gould and Niles Eldredge explain how one of these species might have evolved rather quickly into the other?

ANALYSIS

1. When Darwin went to the Galapagos, he observed fourteen different kinds of finch. One kind had larger and stronger beaks than any mainland finches. What *hypothesis* did Darwin put forward to explain how these finches had developed?
2. In what way, in Darwin's view, were conditions so different on the Galapagos that different kinds of finches could evolve there than would be found on the mainland?
3. When Darwin was looking for *reasons* why birds and tortoises on the

Galapagos were different from those on the mainland, which of the *Ten Principles* was he demonstrating? Explain.

4. On the Galapagos, finches are the only native birds. But these birds live in the same way that other birds live elsewhere: some crack seeds open, others eat insects, and so on. Explain how this observation provides an example of the Principle of Consistency.

5. Explain how Darwin's theory of evolution provides an example of the Principle of Dynamism.

6. While accepting most of Darwin's conclusions, some modern biologists have questioned his view that millions of years would be needed for a major change to come about in an animal or a plant. Explain how this provides an example of the Principle of Tentativeness.

7. Explain how the Principle of Social Limitation made it difficult for many nineteenth-century scientists to accept the theory of evolution.

Questions 8–15 refer to the following account of the work of Lamarck and Weismann:

The idea that one kind of animal might evolve from another was not new with Darwin. In the year of Darwin's birth, 1809, the French naturalist Jean B. Lamarck devised a scheme which explained *how* different characteristics of plants and animals could develop. Lamarck had noticed that in most animals, muscles which are used a lot become large and strong. Those which are not used get smaller and weaker. Similarly, reasoned Lamarck, if a body organ was used a great deal, like the webbed foot of a water bird, it would grow larger and stronger. If an organ was not used, like the wing of a bird which had no need to fly, it would not develop, and might even wither away. Then, said Lamarck, the larger webbed foot or the smaller wing would be passed on to the bird's descendents.

Seventy or so years later, the German biologist August Weismann doubted Lamarck's theory. In Weismann's view, each animal and plant had its own "germ plasm," which today we would call genes. The germ plasm would not be affected by anything that happened during the individual's lifetime. Alter the animal as much as you like, said Weismann, and no change will be seen in its descendants.

To test this idea, Weismann cut off the tails of 1592 mice over 22 generations. He found that even in the 23rd generation, all the mice developed tails just as long as the ones he started with.

8. What was Weismann's *hypothesis*?

9. What *experiment* did he do?

10. What *prediction* must he have made, on the basis of his *hypothesis*, about the result of his *experiment*?

11. What *prediction* would Lamarck have made about the result of Weismann's *experiment* with the mice?

12. What *prediction* would Lamarck have made about the sons of a championship wrestler?

13. What *prediction* would Weismann have made about the sons of the same championship wrestler?

14. In Lamarck's time, the giraffe had just been discovered. Lamarck thought that it would have been descended from an antelope-like animal with a shorter neck. What do you think was Lamarck's *hypothesis* about how the longer neck would have developed?

15. How would Darwin's theory of natural selection explain how a bird which swam in the water would develop strong webbed feet over a period of many generations?

Questions 16 – 20 refer to the investigation by James Lazell, Jr., into the Kahili rock wallaby in Hawaii.

16. What *hypothesis* did this work relate to?

17. What *prediction* was being tested?

18. What evidence in favour of the *hypothesis* was found?

19. What other *hypothesis* did Lazell investigate and then discard?

20. Find a quotation from the article, in the section about Lazell's work, which gives an example of the Principle of Tentativeness.

26 Evolution (II): The Evidence

One of the things that struck Darwin about the Galapagos finches was this: wherever on the mainland some other bird ate a particular kind of food, on the Galapagos there was a finch that ate the same thing. Woodpeckers, for example, probe holes in trees to root out insects; in the Galapagos there are no woodpeckers, but there is a finch that has a similar beak to a woodpecker, which finds and eats insects in the same way.

Marsupials

The same thing is true of the animals in Australia. No true mammals are native to Australia. Rabbits, cats, dogs, and of course sheep are found there now, but they have been introduced there by man. Instead of ordinary mammals (called placental mammals after the structure joining the embryo to the mother's uterus), Australia is populated by marsupials: koala bears, wallabies, kangaroos, and hundreds more. Marsupial young are very small and quite undeveloped at birth. Instead of being nourished through a placenta inside the mother's body until they are more mature, they climb up the mother's fur to a pouch in her abdomen. Inside the pouch are nipples which provide milk. The young live in this pouch until they are fully formed.

In fact, in Australia there are marsupials that resemble many kinds of mammals in the rest of the world. There are marsupial mice, marsupial moles, marsupial cats. There was even, until 1933 when the last one died in a London zoo, a marsupial wolf called the thylacine. Another marsupial, the sugar glider, is a small leaf-eating animal with a parachute of skin connecting its front and rear legs, which allows it to glide great distances — just like the North American flying squirrel.

To cap it all, a sensational discovery was made in 1969 in the caves of Naracoorte, 250 km south of Adelaide. After an hour of crawling through narrow tunnels on hands and knees, explorers found a long low gallery containing perfectly-preserved bones of other marsupials, now long extinct. Among other remains, the explorers found an enormous kangaroo with a neck like a small giraffe, and a giant marsupial the size and shape of a rhinoceros.

Evolution explains these similarities rather readily. If the circumstances (climate, soil, availability of food) are similar in different places, then similar animals evolve by natural selection to take advantage of the opportunities.

Evidence for evolution

This type of evidence, though, is not the only reason that almost all biologists regard the theory of evolution as the best explanation of natural phenomena that has yet been proposed. The theory also explains the arrangement of fossils in the ground, the way in which embryos of animals develop from an egg, the useless bones and other structures in various animals, and the very similar biochemistry shared by most animals and plants. In addition, the theory of evolution fits well with another theory originally proposed by geologists: the theory of continental drift. In fact, the theory of evolution is just as central to biologists as the idea of atoms is to chemists. As scientists, they know that new evidence could turn up that would disprove the theory completely. If that ever happens, their ideas will have to be entirely rearranged.

Continental drift

Australia is not the only place where marsupials exist. There are some — though not nearly as many — in South America, and even in North America, where the opossums are found. In fact, the oldest fossil remains of marsupials have been found, not in Australia, but in South America.

How could marsupials have travelled from one continent to another? The theory of continental drift provides a possible answer. At one time, according to the theory, the continents were not arranged around the world as they are now. All the land on the present continents was joined together in one giant land-mass: Pangaea. This would explain why the same types of fossils — of dinosaurs, for example — more than 225 million years old are similar all over the world. Around 200 million years ago, the land mass split in two: to the north went the land that is now Europe, Asia, and North America. The other continent would have been made up of South America, Antarctica, Africa, and Australia. Antarctica and Australia split off from this southern continent about 165 million years ago.

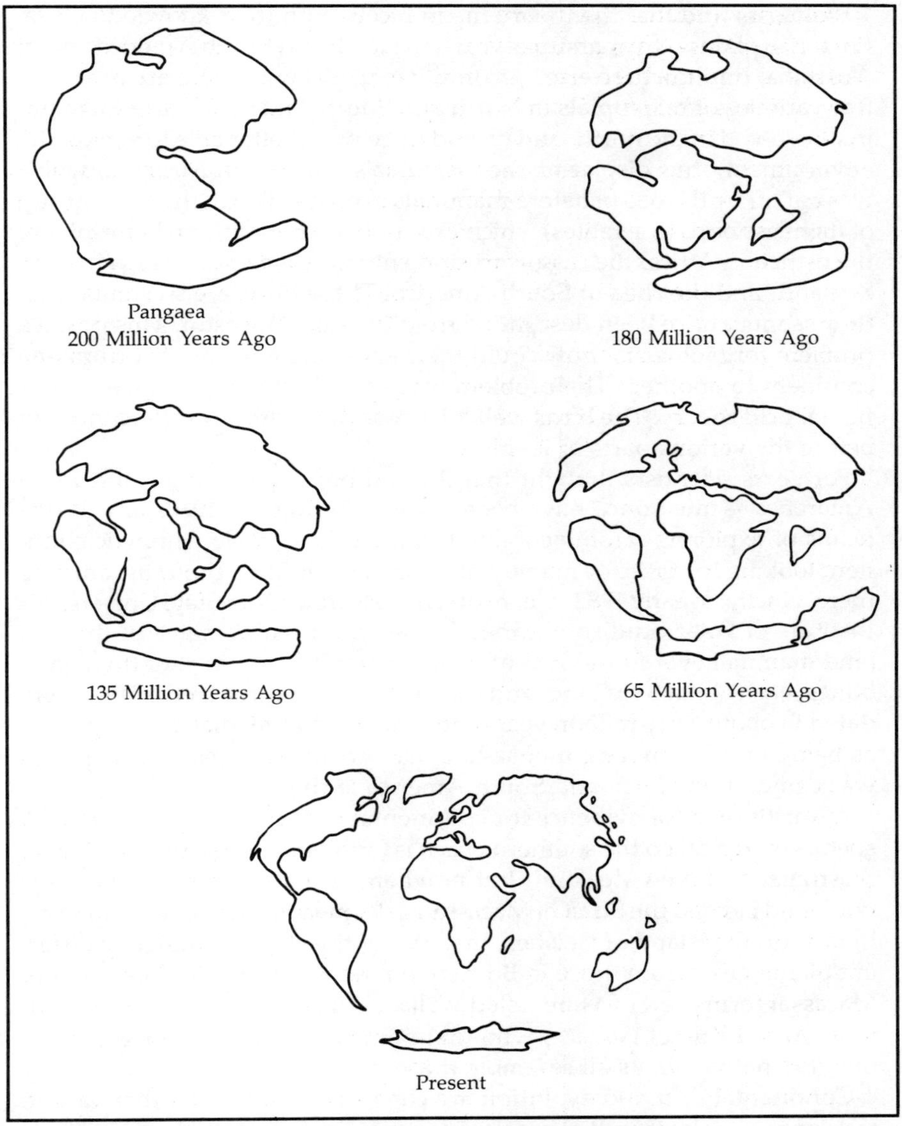

FIGURE 26-1 Pangaea and the Breakup of the Continents

Later still, about 65 million years ago, when the dinosaurs were extinct and mammals were the dominant animals, South America and Africa split apart. Relatively recently, North America separated from Europe, and eventually became joined to South America.

Geologists have strong evidence for continental drift. Rock formations on both sides of the Atlantic Ocean match one another to a remarkable extent. Not only do the rocks match: the orientation of magnetic crystals in the rocks matches as well.

Biologists find that the theory fits in nicely with their knowledge, too. First, it explains why marsupials exist in North and South America and in Australia, but nowhere else. Second, it explains why there are relatively few varieties of marsupials in North and South America. Placental mammals evolved in the north, and by and large were better able to exploit the environment, thus displacing the marsupials — except in Australia, which was cut off by the ocean before mammals evolved. Third, there is a group of flightless birds (the ratites) which exist only in the southern hemisphere: the ostrich in Africa, the cassowary and emu in Australia, the kiwi in New Zealand, and the rhea in South America. These birds are so similar that they seem to have been descended from the same ancestor. This posed a problem for biologists: how could their ancestors have moved from one continent to another? The problem was solved by the geologists' continental drift theory. The birds walked across the southern supercontinent before the various parts of it split up.

For years scientists thought that the southern continents — including Antarctica — must once have been joined. During the 1960s and 1970s, teams of explorers scrambled about in the inhospitable Antarctic continent looking for ratite or marsupial fossils which, they thought, must be there. Finally, in early 1982, a team of scientists from Ohio State University's Institute of Polar Studies returned home with fossil remains of the first land mammal ever found in Antarctica. The find consisted of three jawbones, each about 3 cm long, and each with two or three teeth. They were dated to about forty million years ago, and were immediately recognized as being of a 20 cm long mouse-like marsupial: *polydolops*. This species was common in North and South America at the time.

A fourth piece of evidence for continental drift comes from botany. A species of tree called the southern beech is found in South America, Africa, Australia, and New Zealand, but nowhere else. The southern beech is not found beyond the Strait of Macassar, a deepwater channel in Indonesia. In fact, on the island of Celebes, just east of the strait, completely different plants grow from those in Borneo, on the western side. The Strait of Macassar forms part of a line called Wallace's Line, after the English naturalist Alfred Russel Wallace, who first pointed it out. To the west of the line, the native plants all resemble those in the rest of Asia.

Continental drift and evolution are consistent with each other, and in fact support each other. Biological evidence gives support to the continental drift theory, as we have just seen. And continental drift helps provide evidence for evolution, by answering a number of puzzling questions.

The fossil record

Fossils are the remains of plants and animals preserved in rock formations. As we have seen, fossils provide evidence that the plants and animals which exist today are not always the same as those which existed in the past. Slowly over millions of years, conditions change: land forms, weather patterns, and soil fertility. Animals and plants change too, and adapt themselves to the new environment. Those that cannot change die out. As we have seen, too, fossils can provide evidence of the animals which once inhabited areas like the Antarctic, but which died out when weather conditions changed. Other conclusions from fossil studies also provide important backing for the theory of evolution. Wherever in the world different layers of fossils have been found, the fossils close to the surface have resembled present-day species. The further down paleontologists went, the more likely they were to find species now extinct, and more and more primitive compared with what is found in the world today. The more highly developed and complex animals occur only in the upper layers close to the surface.

One of the most vivid examples of fossil layers occurs in the United States. The Grand Canyon in Arizona is about 1.5 km deep, carved out by the Colorado River. The walls display an amazing variety of fossils, from primitive algae in the bottom layers to relatively advanced reptiles near the top. This progression of layers, though, is not seen only in the Grand Canyon. Paleontologists have found many similar sites all around the world.

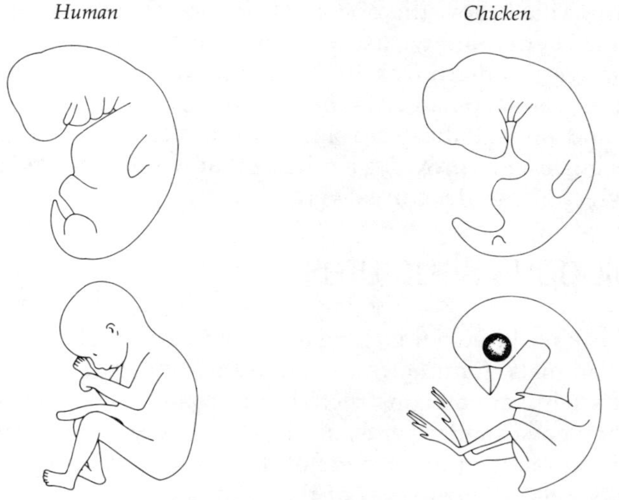

Vertebrate embryos are similar in early stages. Differences become more pronounced as development proceeds.

FIGURE 26-2 Human and Chicken Embryos

Embryology

Something else that many biologists believe supports evolution is the science of embryology: the study of how an animal develops from an egg.

The egg starts out as a single cell, like the most primitive of all living things, the protozoans. From this point, all animals develop along similar lines. At one stage, for example, a human embryo has gill-like structures which in other animals actually do develop into gills. However, in humans and other mammals, they disappear. Later, the human embryo has a tail and body hair like the embryos of all other mammals. In the human embryo, these features disappear before birth, but their presence provides evidence that humans and other animals developed from common ancestors.

Vestigial features

Occasionally, a human baby is born with a distinct tail. Usually, this tail is quietly removed by the doctor. But all of us have a useless bone — the tailbone — which seems to have no function. Biologists regard this tailbone as the remains, or vestige, of the tail that our distant ape-like ancestors in the trees would have found very useful. Humans and other animals have a number of vestigial parts like the tailbone, which now are either useless or harmful. Humans have two ridges that run from the nose to the upper lip. Other animals have much deeper grooves, and they are still useful. In animals like the dog, the groove brings saliva from the mouth to the nose. This is how the dog keeps its nose shiny and damp, which helps it smell better. But we use our noses much less, and do not need to keep them wet. The grooves are of no use to us. Useless extra nipples which an occasional person has, body hair, and the muscles which move the scalp and ears: all these are similar examples of these vestigial structures. Their presence provides evidence that we are descended from animals in which these structures were useful.

Homologous structures

Anatomists have known for a long time about the similar bone structures in various animals. A human arm, a horse's front leg, a dolphin's flipper, and a bat's wing are all constructed in the same way, and from similar bones. They have similar veins, arteries, and nerves. The same observation has been made of other parts of different animals, too. Rear limbs, rib cages, skulls — most parts of the body, in fact — are all remarkably similar in animals of different sizes and shapes, except of course for differences in proportion and size. According to the theory of evolution, these similar, or homologous, structures occur because all the animals are descended from the same ancestors in the distant past.

Biochemistry

Biochemistry is the chemistry of life systems, and our knowledge of it has developed at a great pace in recent times. This understanding, too, has

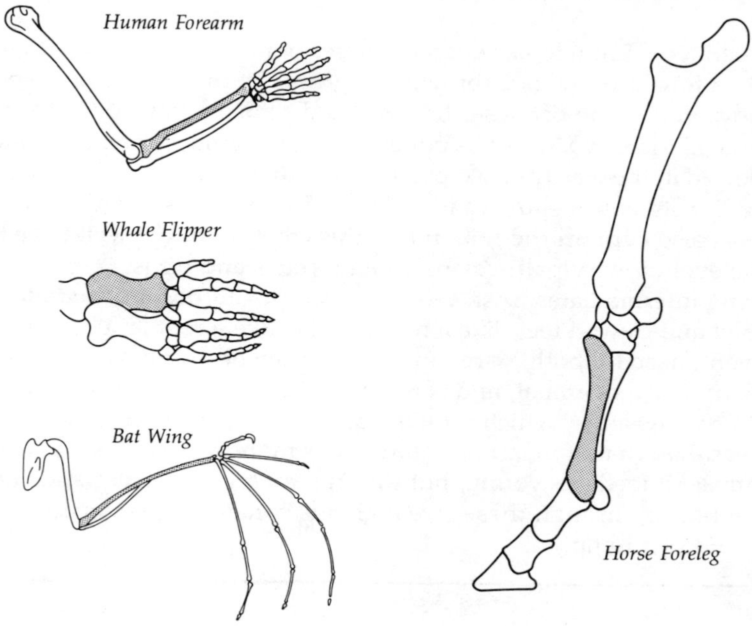

In each case the shaded bone is the radius. (The forelimbs are not drawn to the same scale.)

FIGURE 26-3 Homologous Structures

turned out to fit in well with the theory of evolution. All plants and animals are constructed from proteins made from the same six substances: the amino acids (glycine, alanine, valine, leucine, glutamic acid, and aspartic acid). In the late 1950s it became possible to analyze the remnants of proteins from fossils up to 300 million years old; exactly the same amino acids were found.

More recently, in the 1970s and 1980s a number of other new biochemical techniques became available, such as chromosome banding and protein sequencing. The biochemists who developed these techniques were interested in making comparisons between different animals. They expected, for example, that chromosomes of animals close to each other in evolutionary terms—like fox and dog, or chimpanzee and human—would show up as being very similar in these studies. The chromosomes of mouse and man, on the other hand, ought to show much greater differences. These two animals separated from each other much longer ago, and much further back down the evolutionary tree. When the chromosome studies were done, the result was exactly what had been expected. The closer together the animals were in evolutionary terms, the more similar their chromosomes. Protein sequence studies have shown the same thing, just as had been expected from the theory of evolution: the closer the animals on the evolutionary tree, the more similarly arranged are their proteins.

Intermediate species

Between what seem to be separate classes of plant and animal, there are in fact a few intermediate forms, though not many. The extinct species *archaeopteryx* is one of these. Its fossilized remains have been found in Bavaria, Germany. Most of its bones were very similar to reptile bones in shape and in the way they were joined together. Like a reptile, too, it had teeth. But its bones show that it also had wings. Fossilized remains of feathers also support the notion that this creature might well have been on the evolutionary pathway between reptiles and birds.

Living intermediates exist, too. The platypus, found in Australia, has webbed and clawed feet, like a reptile. Also like a reptile, it has a single rear vent, used for both excretion and reproduction, and it lays eggs. But it is furry like a mammal, and is warm-blooded, so that it can remain active in cold weather, which reptiles cannot do. Its temperature, though, is lower than that of mammals, and varies more widely. Also, like other mammals, it feeds its young, but through enlarged sweat glands rather than nipples. Through these sweat glands comes a liquid which is very thick and rich in fat.

Archaeopteryx: *The presence of wings and feathers defines it as a bird, but the skeleton is quite reptilian.*

FIGURE 26-4 Archaeopteryx

Fossilized remains have been found of a number of other animals midway between one of our present classes of animal and another. Gaps still remain, of course. One such gap is rapidly being filled with fossil specimens: the gap between man and his ape-like ancestors. Paleontologists Richard Leakey, working in Tanzania, and Donald Johansen, working in the remote Afar region of Ethiopia, have discovered numerous remarkably complete fossilized specimens of fairly recent human ancestors of about three million years ago. Debate on the exact course taken by human evolution continues, as does the search for more fossils. But none of the scientists working on these fossils has any real doubt that they are in fact the ancestors of modern man.

QUESTIONS

1. Does the existence of marsupials prove that the theory of evolution is correct? In what way do marsupials support the theory?
2. Outline in point form the evidence for continental drift.
3. Does continental drift prove that the theory of evolution is correct? In what way does the idea of continental drift support the theory of evolution?
4. Does the fossil record prove that the theory of evolution is correct? In what way does the fossil record support the theory of evolution?
5. Does the way in which embryos develop prove that the theory of evolution is correct? In what way does the development of embryos support the theory of evolution?
6. Does the existence of vestigial features and homologous structures prove that the theory of evolution is correct? In what way do vestigial features and homologous structures support the theory of evolution?
7. Does our rapidly-developing biochemical knowledge prove that the theory of evolution is correct? In what way does this knowledge support the theory of evolution?
8. Does the existence of animals such as the archaeopteryx and platypus prove that the theory of evolution is correct? In what way do these animals support the theory of evolution?

ANALYSIS

1. What *prediction* was made by the team of scientists from Ohio State University's Institute of Polar Studies when they went to Antarctica in 1981?
2. Upon what *hypothesis* or *hypotheses* was this prediction founded?
3. The experiment in this case was simply to look for fossils and to examine any that were found. What was the result of this *experiment*?

4. What *prediction* was made by biochemists about the results of their chromosome banding and protein sequencing experiments?
5. What results did these *experiments* produce?

Each of the quotations in questions 6 – 10 provides an example of one or the other of the *Ten Principles*. For each, state which principle is involved, and explain.

6. "The theory also explains the arrangement of fossils in the ground, the way in which embryos of animals develop from an egg, the useless bones and other structures in various animals, and the very similar biochemistry shared by most animals and plants."
7. "As scientists, they knew that new evidence could turn up that would disprove the theory completely."
8. "Animals and plants change too."
9. "This progression of layers, though, is not only seen in the Grand Canyon. Paleontologists have found many similar sites all around the world."
10. "Debate on the exact course taken by human evolution continues, as does the search for more fossils."

27 Down's Syndrome and Radiation: The Work of Dr. Irene Uchida

To many people, a child with birth defects represents a terrible tragedy. To medical researchers, the investigation of the causes for these defects represents a challenge. Only if they understand the causes for the abnormalities can doctors and researchers set about trying to prevent them.

Down's syndrome

One of the most common of these abnormalities is Down's syndrome, first described in 1866 by the English physician John Langdon Down. Physical traits associated with Down's syndrome include short stature, a prominent forehead, a short and broad neck, rough and dry skin, short stubby hands, and a broad single crease across the palm of the hand. The tongue is usually large, which often causes the afflicted person to keep the mouth open most of the time, and sometimes to dribble quite badly. The upper eyelids have an extra fold of skin, which gives a person with Down's syndrome an oriental appearance. This appearance is why the term Mongolism was sometimes used in the past to describe the syndrome.

People with Down's syndrome exhibit varying degrees of mental retardation. In the past, they were normally kept in institutions. Parents were told that special training was needed to cope with and teach these children. Now the trend is the other way. Many parents have found that they can learn to bring up a child with Down's syndrome themselves. Some professional help and some training are necessary, but with love and affection, the children can learn and develop more quickly at home than in an institution. They can be taught to read and write to varying degrees, and can learn to enjoy music and crafts. Many learn to look after themselves

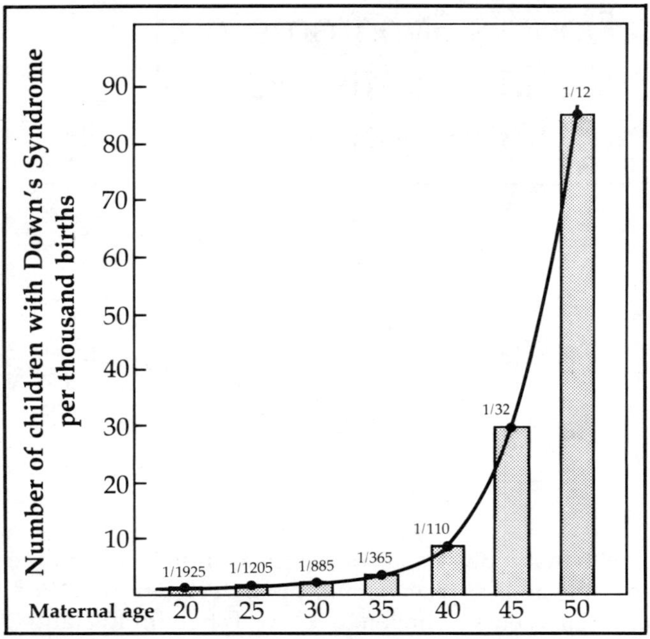

FIGURE 27-1 Histogram representing the chances of bearing a child with Down's Syndrome

most of the time, as long as they get some help. Generally, though, because of the heart defects that are also part of Down's syndrome, these children tend not to live very long. Over half die before they are ten years old. It is rare for a person with Down's syndrome to survive into old age.

About two-thirds of the fetuses who would otherwise be born with Down's syndrome are spontaneously aborted—the mother has a miscarriage before the child is born. Nevertheless, Down's syndrome is the most common of all severe birth abnormalities; about 1 child in 700 is born with it. As a result, many physicians and medical researchers have devoted themselves to investigating this condition. One of the ablest of these researchers is the Canadian geneticist Dr. Irene Uchida. Her investigations are a classic example of detective work in medical research.

The mother's age

In the early 1930s, L.S. Penrose, an English geneticist, pointed out that Down's syndrome children are born much more frequently when the mother is relatively old. One study showed that the probability ranged from 1 in 2000 when the pregnant woman was 20 to over 1 in 12 when she was 49. Why are older mothers more likely to have children with Down's syndrome? Dr. Uchida was confident that there was a reason, and she had an idea about what this reason might be.

One clue came from the cytogeneticists — biologists who specialize in the study of chromosomes. Using techniques which make the cell's chromosomes separate and swell, a French cytogeneticist, Jearoame Lejeune, discovered in 1959 that people with Down's syndrome have an extra chromosome — 47 rather than the normal 46. The chromosome known as number 21 is the one associated with Down's syndrome. Instead of having two of these chromosomes, people with Down's syndrome have three.

Abnormalities in the chromosomes

Dr. Uchida knew about this extra chromosome, of course. She also knew about another study, conducted in the early 1930s on fruit flies. This study showed that female fruit flies exposed to radiation produced a high proportion of offspring with defective chromosomes. Age alone wasn't enough

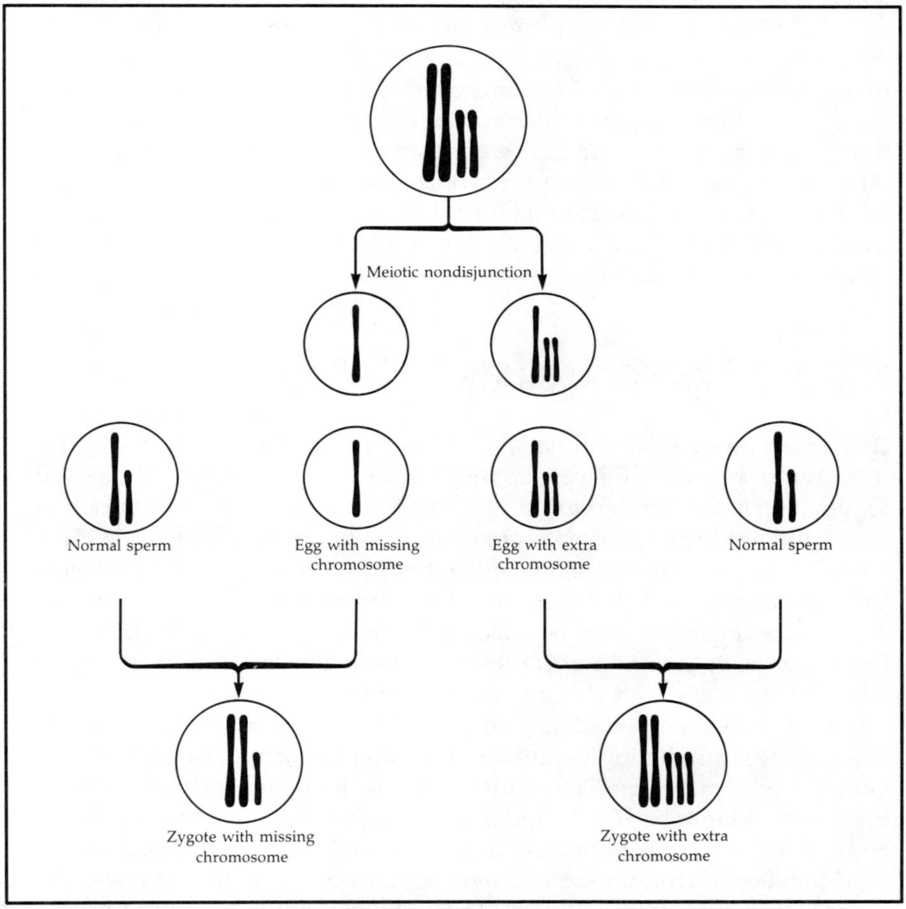

FIGURE 27-2 Meiotic Nondisjunction

—only the older fruit flies which had been exposed to radiation produced the abnormal offspring. To Dr. Uchida, these results suggested that radiation might be responsible for the extra chromosome in people with Down's syndrome. When a woman was exposed to radiation, a change might occur in the chromosomes of her ovaries which would be passed on to a child conceived later on. The older the mother, the more radiation she would probably have been exposed to.

The first step in Dr. Uchida's investigation was to make sure that the 1930s fruit fly study was valid. Working with more modern techniques and equipment, Dr. Uchida and her co-workers repeated it. They obtained the same results as had the earlier researchers: the fruit flies which had been exposed to more radiation produced a high proportion of offspring with defective chromosomes. The defect was one called *nondisjunction*—the failure of one of the chromosomes to separate properly during meiosis. Meiosis is the process in which sex cells (sperm in males, egg cells in females) are produced from ordinary cells. Normally, each sex cell receives only half the number of chromosomes that are found in other cells. In fruit flies, a sperm or egg cell has 4 instead of 8 chromosomes; in humans there are 23 instead of 46. The female fruit flies exposed to radiation had one pair of chromosomes which did not separate in the normal way, so that both members of the pair were transmitted to the offspring. The offspring, therefore, had 9 chromosomes instead of the normal 8. It had inherited the extra one from its mother. As in the earlier investigation, those flies which had not been subjected to the radiation had normal offspring, even though they were just as old as the others.

The first Manitoba study

Perhaps, reasoned Dr. Uchida, Down's syndrome could be caused by radiation too. Just as in the case of the fruit flies, human mothers exposed to radiation before conception could produce egg cells with chromosomes which did not separate properly in meiosis —nondisjunction. Then there would be an extra chromosome, and if this was a number 21 chromosome, the child would have Down's syndrome. Not all the children of a mother exposed to radiation would be affected, because not all cells are damaged. Even if some of the mother's chromosomes had been affected, any child who did not receive an abnormal cell would be normal.

If this idea was correct, then human mothers who had been exposed to heavy doses of radiation would be more likely to produce children with Down's syndrome. In 1960, working from the Children's Hospital in Winnipeg, Manitoba, Dr. Uchida, along with fellow-investigator Elizabeth Curtis, sent questionnaires to, telephoned, and interviewed a hundred families of children with Down's syndrome. If radiation really did cause Down's syndrome, then mothers of these children would turn out to have been exposed to more heavy doses of radiation — abdominal X-ray examinations and fluoroscopies — than mothers of unaffected children.

This is exactly what the results of the study revealed. Many of the mothers of children with Down's syndrome told the investigators about heavy doses of medical radiation, received before conception of the affected child. Far fewer of the mothers from two control groups recalled similar heavy doses. The results were highly suggestive. But, as Dr. Uchida commented at the time, "This finding should be accepted with caution until it can be confirmed by other evidence."

Since the Manitoba study, a number of other similar investigations have been carried out, in the United States, England, Iceland, Japan, and Denmark. In many cases the results were inconclusive — not enough cases were investigated to resolve the question one way or the other. But almost all showed the same tendency for affected children to be born to irradiated mothers. Also, of course, not everyone can remember medical procedures carried out on them years before. All the studies which do show statistically significant differences agree with the Manitoba study. More of the mothers of Down's syndrome children have had heavy doses of radiation than have similar groups of mothers of normal children.

Scientists are always pleased when different studies in different parts of the world point in the same direction. It would be disturbing to find that Down's syndrome is associated with radiation in Manitoba, but not in England.

The second Manitoba study

Studies which require people to respond to interviews and questionnaires always have an element of unreliability. When sent by a busy doctor to hospital laboratories for tests, many people never really understand what these tests are. Even if they do understand them at the time, people sometimes forget what medical procedures were carried out on them years before. So in 1968, Dr. Uchida, working now with research assistants Roberta Holunga and Carolyn Lawler, undertook a different and much more demanding type of study. From the department of radiology at the Winnipeg General Hospital, the three researchers recorded the names of all married women under 60 exposed to abdominal X-ray examinations or fluoroscopies between 1956 and 1959. Then they set out to find these women, and to learn if and when any of them had given birth to a Down's syndrome child. If radiation indeed causes Down's syndrome, then children conceived after radiation would turn out to be more likely to have Down's syndrome than those conceived before exposure.

It was a large and demanding investigation. The researchers searched through the medical records of over six thousand women. Those who had received large doses of radiation had then to be traced, years after the treatment described in the medical records had been carried out. This exhausting task involved, among other things, getting the co-operation of hundreds of doctors, and trying to contact directly the patients who had moved, changed doctors, or who no longer had a family doctor.

From the results of her previous studies, Dr. Uchida expected that mothers would be more likely to have children with Down's syndrome after they had been exposed to the radiation in Winnipeg General Hospital. This turned out to be the case. Children were ten times as likely to be born with Down's syndrome if they were conceived after their mother had been subjected to radiation.

The study was a very large one, and it had to be done with great care. For example, the study involved a comparison between two groups of mothers, one whose children had been conceived before the mothers had received the radiation, and one whose children had been conceived afterwards. The researchers had to make sure that the mothers' ages in the two groups were the same. After all, it was already known that as women became older, they were more likely to give birth to Down's syndrome children. Also, every effort had to be made to locate as many as possible of the 6062 women who had received the large doses of radiation. Otherwise, those women who were easier to find might be different in some way from those who were not found. They might have stabler home lives, for example, eat better food, and keep themselves in better physical shape. It might not seem very likely that this kind of difference would affect the investigation very much, but it is possible. By locating almost all the women concerned, Dr. Uchida and her colleagues greatly reduced this possibility. They therefore could be more confident that the only difference between their groups was the time of exposure to radiation. In terms of experimental design, they went to great lengths to ensure that the experiment was well controlled.

Because these and other necessary precautions were taken, this investigation stands as a model for others of its kind. Dr. Uchida's conclusion, though, was characteristically cautious. In her article describing the study in the prestigious British medical journal *The Lancet*, the word "proof" nowhere occurs. The only claim she made was that the study "presents further evidence that maternal exposure to abdominal X-rays increases the risk of non-disjunction (improper cell division) in subsequent pregnancies."

Work with laboratory mice

These investigations still left a large gap in the Down's syndrome picture. On the one hand were the statistical results from human populations. On the other were the results of studies on fruit fly chromosomes — but the mechanism for meiosis is quite different in fruit flies and humans. To provide a link between these two kinds of study, a direct investigation was needed of the chromosomes of an animal much more similar to humans than fruit flies are — an animal in which meiosis proceeds by the same mechanism as in humans. In 1974, Dr. Uchida embarked upon a completely different kind of study, this time working at McMaster University in Hamilton, Ontario, with fellow-researcher Viola Lee.

Uchida and Lee used a new technique, involving laboratory mice. They counted the number of chromosomes in female oocyte cells — the immature egg cell formed during meiosis. Some of the mice had been exposed to whole-body radiation — not too much, or the chromosomes would be broken. The oocytes of the mice which had been exposed to the radiation were significantly more likely to have an extra chromosome — 21 chromosomes instead of the normal 20. This result represented a further link in the chain of evidence linking radiation, nondisjunction in meiosis, and Down's syndrome.

Dr. Uchida would be the first to agree that Down's syndrome is not always the result of radiation damage to the mother's chromosomes. There are other factors too. About a quarter of Down's syndrome cases, for example, can be traced not to the mother but to the father — chromosomes from the sperm cell, not the egg, fail to separate properly. More recent work has suggested that even when the extra chromosome does come from the mother, radiation is not always the cause. The spindle fibres holding the chromosomes in the oocyte can break down, allowing small chromosomes like number 21 (the one involved in Down's syndrome) to break free. This breakdown occurs more often when the mother is older, and may be caused by toxic chemicals and drugs as well as by radiation.

Scientists are always pleased when research gives clear, simple results which are useful in practical terms, such as advice to doctors to minimize the exposure to radiation of women of childbearing age. But things never seem to turn out as simply as researchers would like. Nevertheless, like most work of its kind, this research points to other promising lines for new investigations.

Directions for the future

The gap in our understanding of the causes of Down's syndrome, though narrower, still remains. As Dr. Uchida comments about her mouse experiments: "Similar experiments are difficult in humans for obvious reasons." But research continues, involving, among other things, artificial cultures of human cells. In early 1984, Dr. Uchida, still at McMaster University, was continuing her work on the problem. Her efforts have already led doctors to be very cautious in administering abdominal X-rays to female patients of childbearing age. The effects of radiation on other chromosomes besides number 21 are also being investigated by studying the chromosomes of miscarried babies and their parents. Dr. Uchida's work also gives valuable guidance to doctors as they decide whether a pregnant woman should undergo amniocentesis — a procedure in which some of the fluid from around the fetus in the amniotic sac is extracted and tested for possible defects in the fetal chromosomes. Amniocentesis can occasionally result in miscarriage or damage to the fetus, and so should not be carried out unless there is some definite reason to suppose that the fetus might be defective.

Dr. Uchida's work, she hopes, will encourage physicians to place "more emphasis on prevention of mental retardation than on treatment and therapy." Her own work has certainly given doctors some important knowledge to help prevent the tragedy of Down's syndrome.

QUESTIONS

1. Down's syndrome is sometimes called Mongolism. Why?
2. In what ways are children with Down's syndrome different in appearance from other children?
3. In the past, children with Down's syndrome were often brought up in institutions. Now, more of these children live at home and are looked after by their parents. List the advantages and disadvantages of having these children stay at home with their parents.
4. Why do people with Down's syndrome tend not to live as long as other people?
5. Write an account of each of the following experiments using the headings: Purpose, Procedure, Results, Conclusion.
   a) Dr. Uchida's fruit fly experiments
   b) Her first Manitoba study
   c) Her second Manitoba study
   d) Her experiments with laboratory mice.
6. From your knowledge of meiosis, or by using a textbook, draw diagrams to show the process of meiosis. Use the diagrams to explain how non-disjunction during meiosis can lead to the offspring having an extra chromosome.
7. Imagine you are a doctor, and a 40-year-old married woman is asking you for advice. She is pregnant and you know she has received large doses of X-rays in a number of previous illnesses. She has heard of amniocentesis and you have to help her decide whether or not to undergo it. What do you say?

ANALYSIS

1. List some of the facts from Dr. Irene Uchida's *data base* when she began the first Manitoba study.
2. What *hypothesis* was being tested in the two Manitoba studies?
3. What *prediction* about the results of the first study followed from the hypothesis? Find a quotation from the article to answer this question.
4. What *prediction* about the result of the second study followed from the hypothesis? Find a quotation from the article to answer this question.

5. What were the main features of the design of the *experiment* in the second Manitoba study that made it more conclusive than the first one?

Questions 6 – 10 concern the following account of Uchida and Lee's McMaster mouse study:

> Uchida and Lee exposed a group of female mice to radiation. Radiation, they thought, might lead to a defect in one of the mouse chromosomes. If this were so, Uchida and Lee expected that they would find an extra chromosome when they examined the oocyte cells of the mice. Normally mice have 20 chromosomes in these immature egg cells. The extra one would have come about because chromosomes damaged by radiation fail to separate properly in meiosis.
>
> Upon examining the mice exposed to radiation, Uchida and Lee found that many of them had 21 chromosomes. This was not particularly surprising, because fruit flies exposed to radiation had previously been shown to produce offspring with an extra chromsome.

6. Find the section of this account which shows the *experiment*.
7. Find the section which shows Uchida and Lee's *prediction* about the results of the experiment.
8. Find the *hypothesis* upon which the prediction was based.
9. Find the *data base* upon which the hypothesis was built.
10. Find the *conclusion* from the experiment, leading to *increased confidence* in the hypothesis.

Questions 11 – 20 present you with quotations from the chapter, each of which provides an example of one of the *Ten Principles*. Identify the principle concerned, and explain how it applies to the quoted sentences.

11. "Down's syndrome is the most common of all severe birth abnormalities. . . . As a result, many physicians and medical researchers have devoted themselves to investigating this condition."
12. "Why are older mothers more likely to have children with Down's syndrome? Dr. Uchida was confident that there was a reason."
13. "The first step . . . was to make sure that the 1930s fruit fly study was valid. . . . Dr. Uchida and her co-workers repeated it."
14. "This finding should be accepted with caution until it can be confirmed by other evidence."
15. "It would be disturbing to find that Down's syndrome is associated with radiation in Manitoba, but not in England."

16. ". . .the word 'proof' nowhere occurs."

17. ". . . Down's syndrome is not always the result of radiation damage to the mother's chromosomes. There are other factors too."

18. "Scientists are always pleased when research gives clear, simple results."

19. ". . .like most work of its kind, this research points to other promising lines for new investigations."

20. ". . . Dr. Uchida comments about her mouse experiments: 'Similar experiments are difficult in humans for obvious reasons.' "

INDEX